Intermediate Offshore Foundations

The late Steve Kay was an independent geotechnical consultant with thirty-three years' experience as a principal engineer with Fugro, and over forty-five years as a geotechnical specialist, mainly in the oil and gas industry, both with contractors and consultants. His expertise was in shallow and intermediate (caisson, bucket, can) foundation design, with extensive worldwide experience in offshore, nearshore and land engineering. He gave suction foundation courses and master classes and wrote the commercially available software package CAISSON_VHM.

Susan Gourvenec is Royal Academy of Engineering Chair in Emerging Technologies in Intelligent & Resilient Ocean Engineering, and Professor of Offshore Geotechnical Engineering at the University of Southampton, UK. Susan is currently Convenor of the International Standardisation Organisation (ISO) committee responsible for developing industry standards for marine soil investigation, marine geophysical investigation and offshore geotechnical design. Susan co-authored *Offshore Geotechnical Engineering* (CRC Press, 2011) and co-edited the proceedings of the inaugural and second *International Symposia on Frontiers in Offshore Geotechnics* (ISFOG).

Elisabeth Palix has eighteen years' experience in offshore geotechnics. She spent twelve years working for Fugro Geoconsulting before joining EDF Renouvelables, where she is working on design and installation aspects of offshore projects. Elisabeth is also a member of the TC 209 (ISSMGE) and has been involved in several geotechnical R&D projects (e.g. SOLCYP, SOLCYP+, PISA, Unified CPT-based methods).

Etienne Alderlieste is a senior geotechnical researcher/consultant for Deltares, where he is working on installation and in-place capacity of intermediate and shallow offshore foundations. Before joining Deltares, Etienne worked as Senior Geotechnical Engineer at SPT Offshore, where he designed suction foundations for the oil, gas and offshore wind industry. He has also installed and reinstalled numerous single-suction anchors and several jacket structures with suction foundations worldwide.

Intermediate Offshore Foundations

Steve Kay, Susan Gourvenec,
Elisabeth Palix and Etienne Alderlieste

CRC Press
Taylor & Francis Group
Boca Raton London New York

CRC Press is an imprint of the
Taylor & Francis Group, an **informa** business

First edition published 2021
by CRC Press
2 Park Square, Milton Park, Abingdon, Oxon, OX14 4RN

and by CRC Press
6000 Broken Sound Parkway NW, Suite 300, Boca Raton, FL 33487-2742

First issued in paperback 2022

© 2021 Steve Kay, Susan Gourvenec, Elisabeth Palix, Etienne Alderlieste
CRC Press is an imprint of Taylor & Francis Group, an Informa business

Publisher's Note
The publisher has gone to great lengths to ensure the quality of this reprint but points out that
some imperfections in the original copies may be apparent.

Visit the Taylor & Francis Web site at
http://www.taylorandfrancis.com

and the CRC Press Web site at
http://www.crcpress.com

British Library Cataloguing-in-Publication Data
A catalogue record for this book is available from the British Library

Library of Congress Cataloging-in-Publication Data
Names: Kay, Steve, author. | Gourvenec, Susan, author. | Palix, Elisabeth,
author. | Alderlieste, Etienne, author.
Title: Intermediate offshore foundations / Steve Kay, Susan Gourvenec,
Elisabeth Palix, Etienne Alderlieste.
Description: First edition. | Abingdon, Oxon ; Boca Raton, FL : CRC Press,
2021. | Includes bibliographical references and index.
Identifiers: LCCN 2021000720 (print) | LCCN 2021000721 (ebook) | ISBN
9781138353534 (hardback) | ISBN 9780367706708 (paperback) | ISBN
9780429423840 (ebook)
Subjects: LCSH: Offshore structures--Foundations.
Classification: LCC TC197 .K39 2021 (print) | LCC TC197 (ebook) | DDC
627/.98--dc23
LC record available at https://lccn.loc.gov/2021000720
LC ebook record available at https://lccn.loc.gov/2021000721

ISBN: 978-1-138-35353-4 (hbk)
ISBN: 978-0-367-70670-8 (pbk)
ISBN: 978-0-429-42384-0 (ebk)

DOI: 10.1201/9780429423840

Typeset in Sabon
by SPi Global, India

Contents

Preface

The origin of this book is as follows. International standard ISO 19901-4:2016 deals with geotechnical and foundation design for the offshore petroleum and natural gas industry. It is due for renewal in 2020. A subgroup, chaired by Susan Gourvenec, was tasked with providing advice (i.e. text, figures and references) on intermediate support foundations. This foundation type is currently not included in ISO 19901-4. In early 2017, since I am now in the knowledge transfer stage of my professional career, and a member of her subgroup, I provided Susan with an 80-page document about suggested advice topics, together with justification for their inclusion. The document had basically four key sections dealing with installation, in-place resistance, in-place response and miscellaneous design considerations. Even if its suggestions were acceptable to the subgroup, it would be inevitable that, because of its length, not all the information would appear in the revised, third edition of ISO 19901-4. Both Susan and I were concerned that much of the material would disappear forever into either the Recycle Bin (MS Windows) or Trash (Mac OS).

Hence this book is basically the above document augmented with material acquired over the past 40-odd years. It is hoped that this book has "plugged" the gap existing between shallow and pile foundations.

The book has two main authors. Due to my ill-health, I provided a good 250-page draft to Susan Gourvenec, who then took on the laborious task of reviewing, editing, improving and getting it into publisher-ready format. In addition, Elisabeth Palix (EDF-EN) kindly contributed sections on intermediate foundations for renewables and weak rock, and Etienne Alderlieste (Deltares, formerly with SPT) was a reviewer and redrew many of the figures ready for publication.

Other acknowledgements are necessary. Firstly, to Ian Smith, my research supervisor, who instilled in me a love of numerical methods in geomechanics, starting with non-linear FEA using 3-noded constant strain triangles. His book programs have been modified to solve the axisymmetric fluid flow and laterally loaded caisson problems presented herein. Secondly, for the transfer of knowledge and practical advice with numerous engineers in many organisations with whom I worked as a civil/geotechnical engineer, later on as an independent geotechnical consultant. Notable organisations include Fugro Engineers BV and SPT Offshore BV. Also, thank you to all CAISSON_VHM users for being sceptical, and not afraid to ask me pertinent questions about the dark corners of in-place resistance and associated intermediate support and anchor foundation topics. Finally, I thank my wife Angela for putting up with my obsession while preparing this book.

Steve Kay
Leiderdorp
The Netherlands
May 2018

Notes from the co-authors

Susan Gourvenec

One afternoon in late 2017, I got a phone call in my office from Steve. He said he would like to write a book on intermediate offshore foundations, based on the materials he had provided to the ISO working group. He wanted the information and the knowledge to be available for engineers to use. He wanted to know if I would be keen to get involved. It seemed like a good idea – so I said yes.

Steve called often to discuss things – always interesting, enlightening and often amusing. Steve retained his humour during all our conversations even in the late stages of his illness, and his desire for 'knowledge transfer' was sincere.

Undoubtedly a talented geotechnical engineer, Steve also struck me as a great enthusiast and citizen of our engineering community. Always generous with his time and knowledge whether at conferences or committee meetings. It is telling that he sent apologies for missing a meeting of the Technical Panel of ISO WG10 just a few days before he passed away. A life cut too short – but Steve certainly made an impression on offshore geotechnics and on the people he worked with and interacted with in his professional life.

I know several geotechnical engineers of my generation, who worked closely with and knew Steve much better than I did. They all speak highly of his generosity in sharing knowledge and of how much they learnt from him. Two such individuals were drafted (by Steve) into developing this book – and I have thoroughly enjoyed the opportunity to work with both Elisabeth and Etienne in creating 'Steve's book', as it has been referred to over the last couple of years as we worked on it.

I would like also to acknowledge the support from the publisher Taylor & Francis, in particular Tony Moore, who enthusiastically took on the book title and worked with Steve and me to put in a proposal. Thanks also to Gabriella Williams, always so understanding about our various delays and missing deadlines, and to Frazer Merritt, who took the book over the finish line to final publication.

Elisabeth Palix

I got the chance to work with Steve at the start of my career on few projects involving suction anchor design. We also got the opportunity to write a few papers together. Steve was a passionate engineer who could give you a call a few weeks after your last meeting to tell you that he had some thoughts about your last discussion. I liked his ability to think out of the box and to take into consideration the opinion of anyone, even a young inexperienced female engineer. I remember that we had to put energy at first to convince some engineers that despite the use of the word pile, anchor piles should not be designed as a pile.

When Steve contacted me to provide him with some support for his book, it was the first I heard about his illness. He was in a hurry to transfer the knowledge he acquired during his career. He was afraid that everything would be lost and disappear into the "Windows recycle bin" (his own words).

This book has been the opportunity to strengthen the links between us at the end of his life. I believe that working on this book gave him some energy to fight his illness and some peace toward the end. I am very glad that I got the chance to be part of this story with Susan and Etienne. We tried together to finish this book as Steve would have done if he was still among us.

Etienne Alderlieste

Not too long after I started working at SPT Offshore, I met Steve. He was contracted to provide engineering support and gave a course on suction foundation design. For another project in 2014, Steve was asked to be involved as reviewer for the detailed design of the suction anchors for an FPSO in the Gulf of Guinea. Throughout the course of that project we discussed a lot of subjects, including engineering and New Zealand. He enjoyed sharing knowledge and simultaneously challenging several design choices, which he was good at, with the aim to end up with a sane design. The final design accounted for potentially detrimental effects of the by-then novel phenomenon known as trenching, without being overly cautious. Not conservative, that term belongs in politics.

During the years that followed we stayed in touch, discussing various engineering matters, personal matters, but also matters concerning CAISSON_VHM – checking functionality of a newer version, assistance when a Windows update messed with things. With Steve's health declining, our contact intensified. He got me involved in this book, and asked me to provide technical support for CAISSON_VHM to assist Richard, who would be taking over.

Susan, Elisabeth and I happily accepted the relatively straightforward task to "finish the book" Steve had drafted. It proved, however, significantly more challenging for all of us than just those three words suggest. A big thanks to Susan and Elisabeth; I really enjoyed our meetings and open discussions!

All co-authors
We would also like to acknowledge, Richard Kay, Steve's son, for seeking out and providing Steve's Mathcad files for recreating figures and analyses presented in the book.

It has been a significant task, but also a great privilege to work on this book. We hope you find it useful.

Susan Gourvenec,
Elisabeth Palix and Etienne Alderlieste
June 2020

If you are interested in using CAISSON_VHM, visit
https://www.casksoftware.com.

Steve Kay
28 November 1946–4 December 2018

Chapter I

Introduction

1.1 INTERMEDIATE FOUNDATIONS

Intermediate foundation describes a foundation geometry and response that falls between "shallow" and "deep" foundations. Shallow foundations are generally taken as having a length to diameter (or breadth) ratio $L/D < 0.5$, while deep foundations (or 'piles') are taken to have $L/D > 10$. Arguably more significant, though, is how the foundation responds under applied loading. Load interaction (between different load components) is significant for intermediate foundations, and they behave as essentially rigid when laterally loaded – distinguishing the response from deep (piled) foundations; while failure mechanisms involve mobilisation of soil strength both deep and shallow – distinguishing the response from shallow foundations.

Intermediate offshore foundations may be broadly divided into the following two types:

- support foundations (point of load transfer at seafloor);
- anchor foundations (point of load transfer usually at depth below seafloor).

The support foundation type includes use for hydrocarbons – platforms, subsea templates, manifolds, wellhead and pipeline supports, riser tower bases; and renewables – wind turbines, transformers and met masts.

The anchor foundation type is mainly used to anchor floating systems or vessels, as a starter pile for pipeline pulls, or to salvage capsized or sunken vessels (including detaching the bow of the Kursk nuclear submarine).

Figures 1.1 and 1.2 give commonly encountered support and anchor foundation examples.

Intermediate foundations have various synonyms, including the terms (in alphabetical order) "anchor pile", "braced caisson", "bucket", "caisson", "can", "free-standing caisson", "monopile", "pile", "skirted foundation", "suction anchor pile", "suction installed skirted foundation", "suction foundation", and "suction pile anchor". A comparison with offshore shallow and deep (pile) foundations is given in Chapter 3.

(a)   (b)

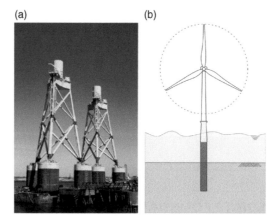

Figure 1.1 Intermediate support foundation type – load applied at seafloor: (a) jacket with suction buckets, Aberdeen Bay offshore wind farm and (b) monopile for offshore wind turbine.

(a)

(b)

Figure 1.2 Intermediate anchor foundation type – load applied beneath seafloor level (at lug): (a) suction anchor (Dijkhuizen et al., 2003) and (b) torpedo pile (De Araujo et al., 2004).

Hogervorst (1980) described the first field trials of full-size suction piles, nearshore Holland, and Senpere and Auvergne (1982) described the first application of suction pile anchors for the Gorm Field, offshore Denmark. Due to their simplicity, reliability and cost savings, suction foundations are also applied to fixed platforms. Bye et al. (1995) described design and installation to support a jacket structure in very dense North Sea sands. Kolk et al. (2001) presented studies for the first lightweight platform (Offshore Cabinda, West Coast of Africa) founded on soft clays and subjected to tensile leg loading. Andersen et al. (2005) presented a keynote paper on the application of deepwater suction anchors, drawn from an industry sponsored study between the American Petroleum Institute (API) and the Deepstar Joint Industry Project VI. The paper identified at least 485 suction caissons had been installed at more than 50 locations in water depths up to 2000 m. These were for oil and gas applications. At the time of writing this book over 40 complete jackets with suction buckets, and more than 500 suction foundation piles, for both oil and gas and renewable platform projects, have been installed worldwide (Tjelta, 2015). In Europe, up to the end of 2019, 4258 monopile foundations have been installed as offshore wind turbine (OWT) substructures, amounting to 80% of the market share by facility (Wind Europe, 2020). Table 1.1 shows that monopiles represent over 80% of all European OWT foundation substructures.

Intermediate foundations, which may be driven, suction installed, vibro- or dynamically installed, have become established for a variety of offshore applications. In many cases they are an economical alternative to piling, with savings on materials, installation equipment and installation time. Perhaps uniquely among foundation designs for fixed structures, intermediate suction foundations offer a realistic option for complete removal. Apart from drilled rock sockets for wind turbines and pylons, there are no intermediate foundation precedents in onshore practice.

Table 1.1 Substructures for Offshore Wind Turbines in Europe (Wind Europe, 2020).

Substructure Type	Cumulative	
	Number	Market Share [%]
Monopile	4258	81.0
Jacket	468	8.9
Gravity Base	301	5.7
Tripod	126	2.4
Tripile	80	1.5
Floating Spar Buoy	6	0.1
Floating Barge	1	0.02
Semi-Submersible	2	0.04
Others	16	0.3

Note(s): 5258 total to 31 December 2019

At the time of writing no (text)book exists that deals solely with intermediate foundations. This book attempts to fill the gap. Aspects of intermediate foundations are covered in the international standard ISO 19901-4:2016 (ISO, 2016a), but the scope is brief and limited to the petroleum and natural gas industries. Design guidance based on existing practice for suction installed intermediate foundations for fixed offshore wind are addressed in the recent Offshore Wind Accelerator guidelines (Carbon Trust, 2019), while new design methods for monopiles for fixed wind is the subject of the PISA project (Byrne, 2017). This book aims to comprehensively address intermediate foundations for both hydrocarbons and renewables drawing from the historical context to current state of the art. The target audience of this book is mainly offshore geotechnical engineers involved in installation and design. In addition, those from other offshore disciplines may be interested in reading selected topics.

1.2 MATCHING MODELS AND DATA QUALITY FOR GOOD DESIGN

A theme throughout this book is that models used should be only as accurate as the data used. In particular for offshore geotechnics, this generally precludes using more advanced models since the required data are simply not obtained during routine offshore investigations, and engineers have to resort to empirical correlations. Examples of missing (or inaccurate) parameter values include (mass) coefficients of consolidation c_v and permeability k, pre-consolidation pressure p'_c, and compression moduli $C_c/(1 + e_0)$, $C_s/(1 + e_0)$. Geotechnical engineers should not be lulled into a false state of security when (for example) a 3D finite element analysis (complete with imagery and animations) has been let loose on poor quality data.

This point – that accuracy of prediction is a function of the quality of the method as well as the data used – was first made by Lambe in his 1973 Rankine lecture (Lambe, 1973). His original figures are reproduced in Figure 1.3. Note that, to state his case, the accuracy contour lines in these figures are either straight or elliptical. Other (intermediate) contour lines are possible. For example, assuming parabolic-shaped accuracy contours (assuming equal weights and consistent with the simple equation "accuracy = model × data"), Figure 1.4 has been prepared. Again, this figure shows that little is to be gained by using more accurate models with low accuracy data. For example, using 50% accurate data and model, 25% (= 50% × 50%) accuracy is obtained. This increases slightly to 40% accuracy (= 50% × 80%) if the model accuracy is increased to 80%. Note that the same 40% accuracy can be achieved if data and model accuracy are both modestly increased to 64%.

A prime example of high-quality offshore predictions being made was for the North Rankin "A" platform, North West Shelf, Australia. In the 1980s,

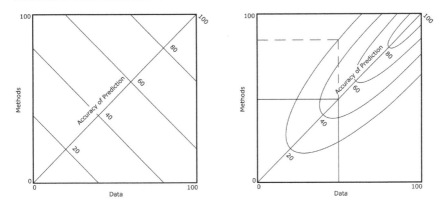

Figure 1.3 Interdependency of accuracy of methods and data on predictions (Lambe, 1973).

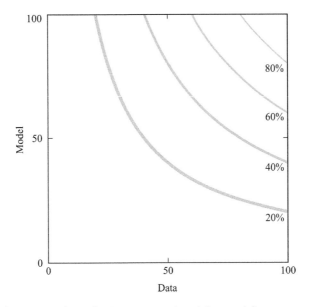

Figure 1.4 Accuracy of predictions – equal weights and "accuracy = model × data."

concerns were raised about the belled pile tip resistance in layered weak calcarenite. High-quality data (triple tube PQ size rock coring providing virtually 100% recovery and detailed core logging, plus stress-path-specific, high-pressure consolidated-drained triaxial testing and in-situ plate load tests) were matched using a double hardening constitutive soil model to perform finite element analyses of single bells and groups of bells (Smith et al., 1988).

1.3 STRUCTURE OF THE BOOK

Chapter 2 sets out definitions of shallow, intermediate and deep foundations, compares characteristics of modes of operation and outlines geometry and terminology for intermediate foundations; Chapter 3 goes on to address determination of appropriate loads for foundation design; Chapter 4 introduces some basics of marine geology; and Chapter 5 considers the interrelationship between loading conditions and soil properties in terms of drainage response. Chapter 6 presents hazards, uncertainties and risk minimisation illustrated with case histories. Chapter 7 presents the process and components of geotechnical design for intermediate foundations from desk study to site investigation. Chapter 8 outlines the design basis for intermediate foundations in terms of installation, in-place resistance and in-place response. Chapters 9–11 present detailed discussion about these three key topics. At the time of writing, there is no general consensus as to which design methods/models are appropriate for intermediate offshore foundation installation (Chapter 9), in-place resistance (Chapter 10) and in-place response (Chapter 11). Hence, there is comprehensive discussion and advice on models to be found in these three key chapters. Finally, in Chapter 12, some 'miscellaneous' design considerations are presented.

Chapter 2

Offshore foundation types and mode of operation

2.1 DEFINITIONS – SHALLOW, INTERMEDIATE AND DEEP FOUNDATIONS

Table 2.1 gives definitions for "shallow", "intermediate" and "deep" foundations. "Shallow" foundations are mostly for lightly loaded seafloor structures, while heavily loaded fixed platforms are founded mainly on "deep" (pile) foundations – with the exception of platform gravity-based structures (GBS). A third foundation type is "intermediate", situated between shallow and deep. Examples include small stubby caissons to support seafloor structures, monopiles for offshore wind turbines, and caisson anchors for semi-submersible or other types of floating structures.

Intermediate foundation installation and in-place response (resistance and stiffness) *both* need to be considered, usually concurrently. Intermediate anchor resistance in normally consolidated clay is a function of time after installation, while an appropriate Factor of Safety (FOS) on in-place resistance has to be assigned *a priori* in order to determine the foundation geometry. Design requires simultaneous input from (and interaction between) geotechnical and structural engineers. These aspects of intermediate foundation design differ from those normally used to design shallow or deep pile offshore foundations.

Intermediate foundations must be installed to their design penetration below seafloor. For intermediate support foundations, this ensures sufficient in-place resistance to withstand combined vertical, horizontal, moment and twist (VHMT) loads. Unlike piles, proof of sufficient V resistance during installation is not necessarily proof of sufficient resistance under VHMT loads. This is due to interaction effects: the available axial V resistance decreases as H and/or M and/or T load increase. Similarly, anchors need lug levels at the correct depth below seafloor to optimise pullout resistance and ensure a near horizontal translation mode.

Table 2.1 Offshore Foundation Definitions.

Foundation Type	Approximate Embedment Length: Diameter Ratio, L/D [–]	Resistance[a]/Load Transfer/ Response[b] Characteristics due to Seafloor VHMT Loads[c]	Foundation Examples
"shallow"	L/D < 0.5	Resistance is a function of VHMT loads, i.e. coupled. Majority of VHMT load is transferred to embedment depth. Similar responses at seafloor and embedment depth.	Pipeline and cable infrastructure, gravity base structures, jacket mudmats, both skirted and unskirted.
"intermediate"	$0.5 \leq L/D \leq 10$	Resistance is a function of VHMT loads, i.e. coupled. Part of VHMT load is transferred to embedment depth. Lateral response is essentially rigid under HM load, with ≈ constant rotation θ_{xz} with depth. Responses differ at seafloor and embedment depth.	OWT monopiles, caissons and anchor piles (installed by deadweight, followed by either suction assistance, impact driving or vibratory driving).
"deep"	L/D > 10	VT resistance is a function of VT loads, HM resistance[d] is a function of H load only, i.e. uncoupled axial/twist and lateral resistance. Some of V load[e,f], but little or no HMT load, is transferred to embedment depth. Some V response, but negligible HMT response, at embedment depth.	Pipe piles for fixed platform or anchor piles for floating platforms.

[a] Resistance: ultimate under combined axial, lateral, moment and twist (VHMT) loads at seafloor
[b] Response: forces (including moment M and torsion T) and displacements (including rotation θ_r and twist θ_t)
[c] Seafloor VHMT loads: see Figure 2.2a
[d] Second order effects ignored (e.g. p-delta for laterally loaded piles)
[e] Ratio of seafloor V load: tip V resistance can be almost zero for a pile in clay, but up to 50% for a pile end-bearing in sand
[f] Flexible foundation: soil system assumed (Murff, 1975)

Table 2.2 Foundation In-place Resistance Assessment Models/Software.

Foundation Type	Modes	References/Software
Shallow Foundations (embedment ratio L/D < 0.5) Figure 2.1	Effective area concept Resistance envelope	ISO 19901-4 (2016a) API RP 2GEO (2011) ISO 19901-4 (2016a) API RP 2GEO (2011) Feng et al. (2014) Gourvenec et al. (2017)
Intermediate Foundations (0.5 ≤ L/D ≤ 10) Figure 2.2 (clay) Figure 2.3 (sand)	Rotational failure	CANCAP2 (Fugro, 2009) HVMCAP (Norwegian Geotechnical Institute, 2004) CAISSON (Kennedy et al., 2013)
	Zero rotation failure	DNV RP-E303 (2017) API RP 2SK (2005) AGSPANC (AG, 2003) FALL16 (OTRC, 2008)
	Resistance Envelope	Vulpe et al. (2014) CAISSON_VHM (Kay and Palix, 2011; Kay, 2015)
Deep (Pile) Foundations (L/D > 10) Figure 2.4	Axial Lateral	ISO 19901-4 (2016a) API RP 2GEO (2011) OPILE (Cathie Associates, 2014)

2.2 MODES – SHALLOW, INTERMEDIATE AND DEEP FOUNDATIONS

Mode of failure is one method of comparing different responses of shallow, intermediate or deep foundations. In this section, different modes of in-place resistance are illustrated for each foundation type to indicate the variety of response within each foundation type and key differences between foundation type. Detailed considerations for in-place resistance, along with installation and in-place response (i.e. stiffness), are covered in Chapters 9–11.

Table 2.2 summarises modes and software (both proprietary and commercial) for assessing in-place resistance of offshore shallow, intermediate and deep foundations for undrained soil response.

The following paragraphs summarise and compare modes for the three foundation types and show that, like shallow foundations, VHM load interaction and soil layering should be considered for intermediate foundations.

Shallow Foundations (embedment ratio L/D < 0.5), Figure 2.1:

There is a wide variety of possible failure modes and associated models, each with its own assumptions and limitations. For example, for an embedded shallow foundation or jack-up spudcan under pure V load,

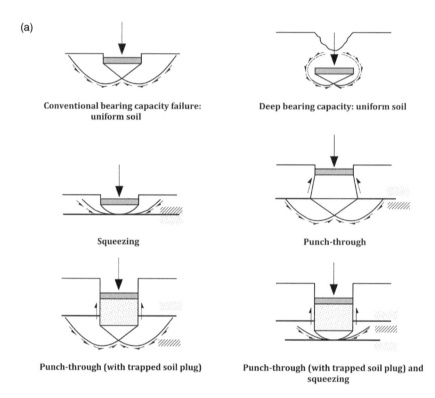

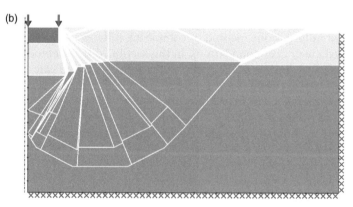

Figure 2.1 In-place failure models for shallow foundations (0 ≤ L/D < 0.5): (a) general shear, squeezing and punch-through failure modes under pure V load (ISO 19905-01:2016b), (b) punch-through sand into clay failure mode under VH load (Ballard et al., 2010). (Permission to reproduce extracts from 19901-4 for a is granted by BSI.)

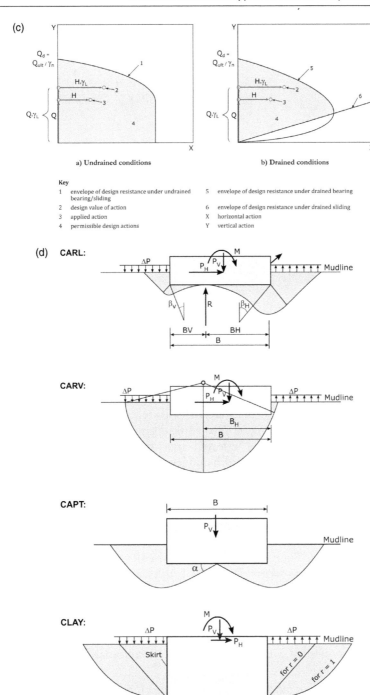

Figure 2.1 Continued: In-place failure models for shallow foundations ($0 \leq$ L/D < 0.5): (c) VH resistance envelopes for sand and clay (ISO 19901-4:2016a), (d) rotational failure under VHM loading. (Permission to reproduce extracts from 19901-4 for c is granted by BSI.)

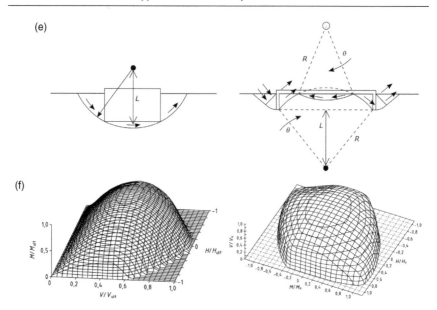

Figure 2.1 Continued: In-place failure models for shallow foundations ($0 \leq$ L/D $<$ 0.5): (e) internal scoop mechanisms (Bransby and Yun, 2009), (f) VHM yield envelopes for surface foundations with and without tension (Taiebat and Carter, 2002).

there are six axisymmetric failure modes (Figure 2.1a). The modes, and models to capture them, become more complicated for combined VH(M) loading, especially for multi-layered soils (e.g. Figure 2.1b), and are routinely simplified (e.g. Figure 2.1c). A traditional simplification for shallow foundations is to use the "effective area" method (ISO 19901-4: 2016a). The effective foundation width and lengths are given by $B' = B - 2\, e_x$ and $L' = L = 2\, e_y$ where $e_x = M_y/V$ and $e_y = M_x/V$. The point of V load application must lie in the middle third in order to avoid tensile stresses at foundation embedment level. It is cautious, especially for embedment ratios of around 0.5. In addition, plane strain vertical slices may be considered, usually with shear contributions on the two out-of-plane faces to account for 3D effects (Figure 2.1d). More possibilities occur if the foundation is not solid but has 'skirts' (Figure 2.1e). A variety of VHM(T) resistance envelopes are available for rectangular and circular foundations on undrained soil (Figure 2.1f). Unlike traditional bearing capacity methods, which use linear load inclination factors, these use 3D FEA to derive VHM(T) interaction effects more rigorously. Also, the "effective area" is no longer needed, since the M resistance is automatically obtained for a given VH load.

Intermediate Foundations (embedment ratio $0.5 \leq$ L/D $\leq$ 10), Figure 2.2 (clay) and Figure 2.3 (sand):

All the modes shown in Figure 2.2 are for rigid foundations in undrained soil and under combined VHM load. The first three modes use 2D plane

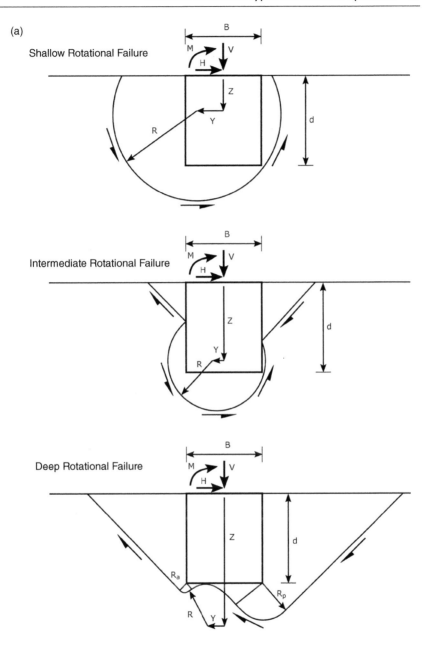

Figure 2.2 In-place failure modes for intermediate foundations (0.5 ≤ L/D ≤ 10) in undrained soil: (a) "shallow, intermediate and deep" caisson rotational failure (Kolk et al., 2001).

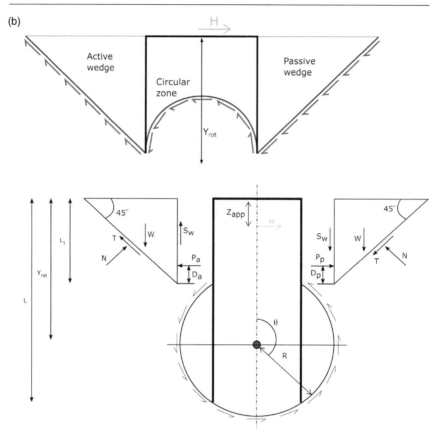

Figure 2.2 Continued: In-place failure models for intermediate foundations (0.5 ≤ L/D ≤ 10) in undrained soil: (b) rotational mechanisms (Kennedy et al., 2013).

strain limit equilibrium and shear contributions on the out-of-plane faces, a variety of rotational failure modes, and options for pile head fixity and gapping behind the pile. The fourth (Murff and Hamilton, 1993) uses upper bound plasticity theory and models the soil as a conical soil wedge with horizontal plane strain flow around the pile at depth. Figure 2.2e shows a "tongue"-shaped VHM envelope developed from ≈ 5500 finite element analyses (FEA). More resistance envelope models are given in Section 10.12.2 (Undrained Soil Response) and Table 10.6. Note that, like shallow foundations, these resistance envelope models permit foundation optimisation without resorting to 3D FEA. Finally, Figure 2.2f shows a failure model specifically for anchor chain trenching.

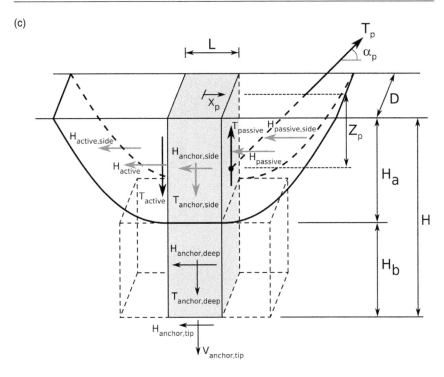

Figure 2.2 Continued: In-place failure models for intermediate foundations (0.5 ≤ L/D ≤ 10) in undrained soil: (c) "fixed-head" anchor pile under chain T load (after Andersen and Jostad, 1999).

There are fewer models for sand than for clay, and these are shown in Figure 2.3. The first (Broms, 1964a, 1964b) is for laterally loaded caissons and the second shows the Butterfield and Gottardi (1994) "rugby-ball"-shaped VHM envelope for shallow (not intermediate) foundations.

Deep Foundations L/D > 10), Figure 2.4:

Unlike for shallow and intermediate foundations, the axial and lateral failure models for deep pile foundations are uncoupled. This is reasonable: uncoupling accounts for the fact that the soil fails first during axial loading whereas the pile fails first (bending stress) under lateral load. There are no equivalent VHM(T) resistance envelopes for deep piles – only axial resistance at pile head, V_{max} versus depth profiles for various pile diameters D.

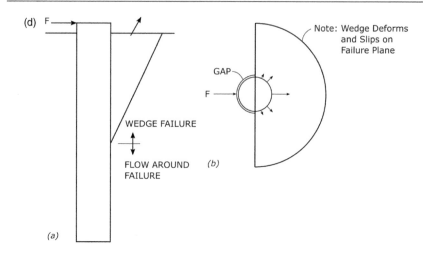

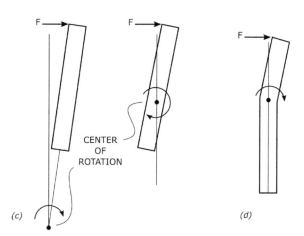

Figure 2.2 Continued: In-place failure models for intermediate foundations (0.5 ≤ L/D ≤ 10) in undrained soil: (d) pile under VHM load (Murff and Hamilton, 1993). (With permission from ASCE).

Axially loaded open-ended pipe pile compressive resistance is the sum of skin friction and end-bearing, and there are two possibilities. As shown in Figure 2.4a, essentially failure occurs either in a "coring" or "plugging" mode, and the minimum resistance value is governing. The remaining three figures deal with lateral resistance modes; the first two show a single and double plastic hinge occurring for surface HM load and anchor pile T_{lug} load respectively (Figure 2.4b and 2.4c).

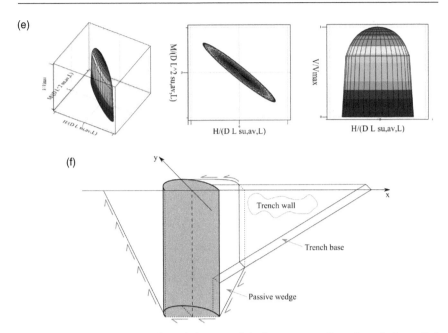

Figure 2.2 Continued: In-place failure models for intermediate foundations (0.5 ≤ L/D ≤ 10) in undrained soil: (e) VHM envelope for clay (Kay, 2015), and (f) anchor pile chain trenching failure model (Alderlieste et al., 2016).

Figures 2.4b and c are for clay profiles. In sand, earth pressures are slightly more difficult to compute and are usually assessed using passive wedge theory (Figure 2.4d).

Note that:

- Intermediate foundations behave like piles when axially loaded – i.e. they usually either respond "plugged" or "coring". However, unlike piles, there is a third mode ("leaking") which may occur during pull-out, more details of which are given in Section 10.5 (Maximum Axial Resistance).
- Modes become simpler and less numerous as one progresses from "shallow" through "intermediate" to "deep" foundations. This is because shallow foundation resistance is extremely sensitive to small variations in soil layering and strength. On the other hand, deep foundations obtain their resistance by averaging shear strength over larger areas, and are therefore not so sensitive to weaker layers.

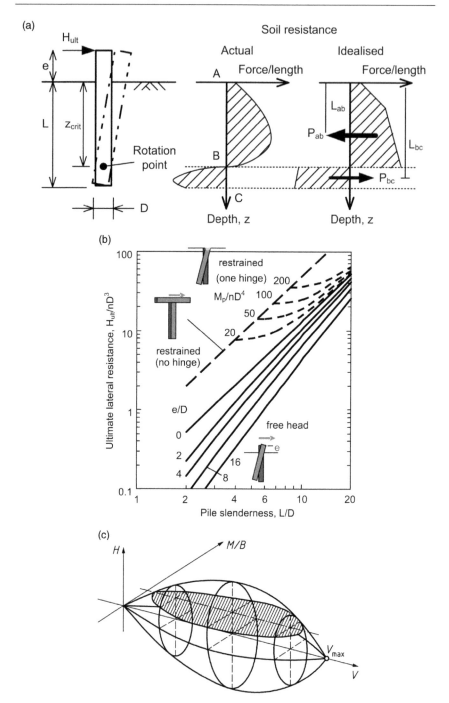

Figure 2.3 In-place failure models for intermediate foundations in drained soil: (a) HM load on laterally loaded rigid pile (Randolph and Gourvenec, 2011) and (b) VHM envelope for shallow foundations (Butterfield and Gottardi, 1994).

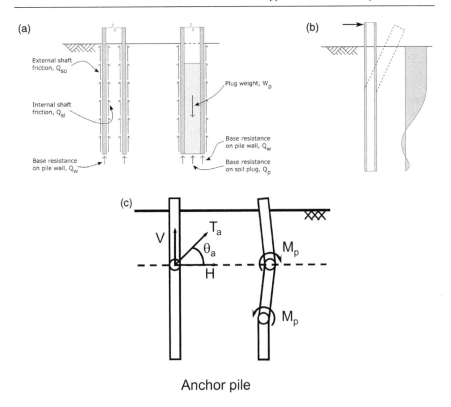

Anchor pile

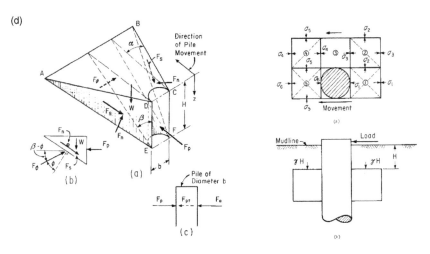

Figure 2.4 In-place failure models for deep (pile) foundations (L/D > 10): (a) axial (soil) failure, (b) and (c) lateral (pile steel) failure, (d) lateral (soil wedge) failure (Reese et al., 1974).

2.3 INTERMEDIATE FOUNDATION GEOMETRY

2.3.1 General

Intermediate *suction* foundations are closed at the top, open at the base, and generally cylindrical. In clay, shapes other than cylindrical may be feasible (for example, square or triangular).

Suction foundation top plates, which contain vent(s), are usually flat. Top plates may be domed in order to reduce steel bending stresses should extremely high underpressures be anticipated. Domed top plates have a single central vent. A geotechnical side benefit is that domes can accommodate more soil plug heave. If inverted T-beam stiffeners are placed below the flat top plate, "rat-holes" are generally added to facilitate water egress. Apart from depth markers on the outer surface (used for installation monitoring), all steel cylinder surfaces are left unprotected in order to maximise available friction resistance. If the surfaces are not rusty, or are painted or treated in other ways that reduce skin friction, reduced interface friction factors should be considered. Suction foundation embedment ratios are a function of foundation type and soil conditions, and typically vary between 0.5 and 6. Cylinder wall thicknesses are much less than driven pipe piles, typically $D/250$ to $D/100$. Top plate thicknesses are higher than cylinder wall thicknesses, usually between $D/100$ and $D/50$.

With the increase of turbine size, monopiles for offshore wind farms are experiencing an increase in diameter, but similar lengths, leading to a decrease of their slenderness ratio (L/D). In the last decade or so, turbine capacity increased from 4 to 12 MW and turbine suppliers have already announced that 14 MW turbine will be on the market soon. Current projects consider monopile diameters in the range of 7 to 10 m. Due to the conjunction of increasing turbine size and the growing maturity of this industry (i.e. better optimisation to be linked with a significant R and D effort in the past few years), the associated embedment ratios are decreasing from a L/D of 5 to L/D of 3 (or even less). Limits on the size of monopiles are likely to be driven mainly by the manufacturer's limits in terms of fabrication and transport (in Europe some manufacturers can fabricate a monopile 12 m diameter and up to 120 m long and 2000t) or the offshore installation vessels available on the market.

2.3.2 Internal stiffeners

Internal stiffeners may be added to intermediate foundations. Examples include:

- Full-height cross plates for support foundations – to either correct tilt during installation by using unequal compartment underpressures, or to reduce soil plug primary consolidation settlement in normally consolidated clay profiles.

- Full- or part-height cross plates for support foundations – to prevent against an internal scoop mechanisms in caissons with embedment ratios < 2 in non-uniform soil.
- Ring stiffener at lug level for anchor foundations – to distribute chain loads and reduce structural steel stresses.
- Ring stiffeners along the length of the caisson – to improve pile bending stiffness and reduce internal friction during penetration.

2.3.3 External stiffeners

External stiffeners (usually two vertical beams) may also be added to the outside of the cylinder to reduce structural steel stresses for foundations installed by skidding (instead of lifting by crane). External stiffeners added above the top plate are not usually of interest to geotechnical engineers.

The cylinder is usually of constant outside diameter and mainly of uniform wall thickness. However, in sand profiles, the wall (and stiffener) thickness may increase at or near tip level; these friction breakers reduce internal friction. Similarly, for anchor foundations, the wall thickness may increase at or near lug level. In clays, the use of vertical stiffeners is preferred over ring stiffeners to reduce the possibility of gapping (vertical water-filled annuli or slots) and twist (misalignment in plan).

2.3.4 Protuberances

Protuberances include mooring and lifting padeyes/lugs, longitudinal or ring stiffeners, changes in wall thickness, mooring chain, launching skids, water injection tubing and others. Occasionally, fins may be added to increase lateral resistance of intermediate anchor foundations. Fins are two steel plates (with stiffeners to reduce bending stresses) sticking out from the caisson wall orthogonal to the H load direction.

2.3.5 Embedment ratio

Generally the objective is to get the foundation as deep as possible in order to maximise foundation resistance. This is because soil generally becomes stronger (and hence unit values of in-place resistance increase) with depth. However, suction installation (not in-place resistance) considerations are usually governing. If an infinitely large underpressure can be applied, and the steel cannot yield, then installation stop criteria are usually soil plug liquefaction (sand) and soil base failure (clay). However, intermediate foundations installed by impact driving have the same stop criteria as deep (pipe) piles – steel fatigue and maximum blow count.

For monopile foundations for offshore wind farms, the diameter is generally governed by frequency and fatigue requirements. The embedment is
adjusted to limit the accumulated displacement at the top and to ensure that
a small variation of the length would not induce a significant variation of the
mudline displacement. There is no specific installation constraint that would
limit the embedment ratio apart from site specific stratigraphy (e.g. presence
of relatively shallow bed-rock). As discussed in Section 2.3.1, L/D ratio tends
to decrease with increase of turbine size, leading to more rigid foundations.

Figure 2.5 shows that (unlike suction support and anchor foundations),
offshore wind turbine monopile foundations are likely to increase in diameter D (and decrease embedment ratio L/D) in the future.

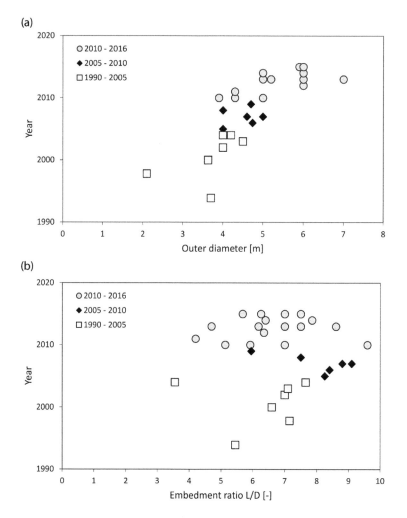

Figure 2.5 European monopile geometry trends. Geometry data from Negro
et al. (2017); wind farm inauguration dates from www.4coffshore.com.

2.4 SUMMARY OF INTERMEDIATE FOUNDATION TERMS

Intermediate foundation: foundations with an embedment ratio between 0.5 and 10. Usually circular in plan. Foundation behaves essentially rigid when laterally loaded. See Section 3.1.

Intermediate suction foundation: steel cylinders which are (temporarily or permanently) closed at the top by a steel top plate and installed using underpressure ("active suction") and/or water pressure less than ambient pressure is generated inside the cylinder under the top plate during tensile loading ("passive suction") and possibly removed/extracted using overpressure.

Underpressure: water pressure inside the cylinder under the top plate which is less than ambient pressure.

Overpressure: water pressure under the top plate which exceeds ambient pressure inside the cylinder.

Both pressures are relative to hydrostatic at seafloor (not under the foundation top plate).

Installation: launching or lowering the foundation into place on the seafloor, followed by further lowering using self-weight and suction assistance to its design penetration depth below seafloor.

Retrieval: foundation recovery as a contingency measure during installation, and subsequent re-installation.

Removal: permanent recovery of foundation after completion of the operational phase. Unlike retrieval, this may include raising it back to the installation vessel.

Base failure: reverse end-bearing failure at foundation tip level (when using underpressure, clay).

Liquefaction: zero effective stress in soil plug (underpressure, sand).

Piping: hydraulic leak between water plug under the top plate and seafloor around the foundation tip (under pressure, sand).

Plug heave: soil plug surface heave (both clay and sand).

Set-up: inner and outer friction increase with time after installation (normally consolidated and lightly over consolidated clay, and loose to very dense sands).

Tilt: foundation non-verticality or out of plumb.

Twist: foundation misalignment or misorientation in plan.

Stick-up: foundation height above seafloor to allow for general seafloor slope, local seabed variations and steel displaced in soil, and other effects.

Embedment ratio: foundation embedment depth length divided by the shortest plan dimension, equal to L/D for circular foundations.

Chapter 3

Loads

3.1 INTRODUCTION

Loads are the starting point for any foundation design and installation assessment. This is because a lightly loaded structure (for example, a 100 kN compressor) can be designed with a different foundation type than a heavily loaded structure (for example, an offshore wind turbine monopile subjected to significant seafloor lateral loads, H = 2 MN, M = 150 MNm).

In addition, offshore structural and geotechnical engineers have different perspectives about loads. Table 3.1 shows that for support foundations: (a) only the most critical (lowest factor of safety, FOS) load cases need geotechnical analysis; (b) structural and geotechnical load formats are different; (c) geotechnical engineers often assume MH loads are generally co-planar, although less so for offshore wind, and (d) structural engineers usually consider only in-place conditions. For anchor foundations, the interface is simpler and perspectives are broadly the same – there are fewer load cases, and structural loads have now only 2 degrees of freedom (DOF) instead of 6.

3.2 UNITS, SIGN CONVENTIONS AND REFERENCE POINT

Offshore engineering practice is to use the following SI system:
   kN, m, tonne mass, seconds and degrees

3.2.1 Example – units

Note that using tonne mass eliminates the need for including gravity (g = 9.81 m/s^2) in dynamics problems. Three examples neatly illustrate this.

Table 3.1 Structural and Geotechnical Perspectives of Loads.

	Perspective	
Consideration	*Structural Engineer*	*Geotechnical Engineer*
Support Foundations (point of load transfer at seafloor)		
Software	SACS, Nastran, SESAM, ANSYS, etc.	Analytical, PLAXIS 3D, ABAQUS, etc.
Number of Load Cases	Many (FLS, ULS, ALS, dynamic, seismic).	Fewer (ULS, SLS, ALS, seismic).
Analysis Type	In place response only. Usually linear (due to high # load cases), non-linear only for a limited number of load cases.	Routine projects – usually ULS (installation and in-place resistance) and linear (in-place response) and stiffness response (both non-linear and linear)
Foundation Loads	F_x, F_y, F_z, M_x, M_y, M_z (6 DOF)	V, H, M, T (4 DOF) as well as Fx, Fy, Fz, Mx, My, Mz (6 DOF)
Foundation Displacements	δ_x, δ_y, δ_z, φ_x, θ_y, Ψ_z (6 DOF)	δ_z, δ_x, θ_{xz}, Ψ_{xy} (4 DOF)) as well as δx, δy, δz, φx, θy, Ψz (6 DOF)
Miscellaneous	Foundation Stiffness	Foundation Settlements
Anchor Foundations (point of load transfer usually at depth below seafloor)		
Software	None for foundation steel. Mooring systems: Orcaflex, SESAM, etc., for chain loads at seafloor.	Analytical, PLAXIS 3D, ABAQUS etc.
Number of Load Cases	Few, usually either two (Intact condition and redundancy check – ISO 19901-7) or three (All lines intact, One line damaged, Transient – O&G Organisations), possibly assess MBL.	Same number as Structural
Analysis Type	None – chain loads generally provided at seafloor.	Anchor chain inverse catenary in soil ULS (installation and in-place resistance)
Foundation Loads	Seafloor: $T_{seafloor}$, $\theta_{seafloor}$ Lug level: T_{lug}, θ_{lug} (= V_{lug} and H_{lug}, 2 DOF)	V_{lug}, H_{lug} (= VHM at seafloor)
Foundation Displacements	Not considered.	Not considered.
Miscellaneous	FEA lug stiffener stresses.	In-place tilt and twist.

Note(s): Displacement symbols φ, θ and Ψ are consistent with those commonly used for ship motions: x = surge, y = sway, z = heave
φ (phi) = roll, θ (theta) = pitch, Ψ (psi) = yaw

Example 1: V_p, the compression wave speed in an elastic solid, is given by $V_p = \sqrt{(E/\rho)}$. For steel, Young's modulus E = 210e6 kPa and mass density ρ = 7.7 tonnes/m³, giving V_p = 5200 m/s, which is the correct speed. No g value has been used.

Example 2: The first natural frequency of a lumped spring-mass SDOF (single degree of freedom) system is given by the equation $\omega_{n1} = \sqrt{(k/m)}$, where ω_{n1} is frequency (radians/second), k is spring stiffness (kN/m) and m is mass (tonnes). Again g = 9.81 m/s² is absent in the equation.

Example 3: One newton (1 N) is defined as the force required to accelerate an object with a mass of 1 kilogram (1 kg) by 1 m/s². Using Newton's Second Law (F = ma), we have acceleration a = F/m, Hence unit acceleration a_1 = 1 N/1 kg. Multiplying the top and bottom by 1000, we now have the same a_1 value, i.e. a_1 = 1000 N/1000 kg. Hence, using kN force and tonne mass automatically gives correct acceleration units (m/s²), making conversion factors unnecessary.

Geotechnical engineers should request/obtain/report centric loads at seafloor. This is the point of load transfer; see Figure 3.1a. This is important for both shallow and intermediate support foundations under combined (VHMT) loading. Centric means at the foundation geometric centre line. This is not necessarily the point where the platform leg meets the foundation – they could be offset. A typical reason for the offset is that platform installation barges have a fixed width. Offsetting platform legs gives a few more metres of clear spacing between the caissons. Another reason for offsetting legs could be for optimisation of load transfer to the soil. For anchor foundations, chain loads (and inclination angles) are usually given at seafloor, not at anchor lug (padeye) level; see Figure 3.1b.

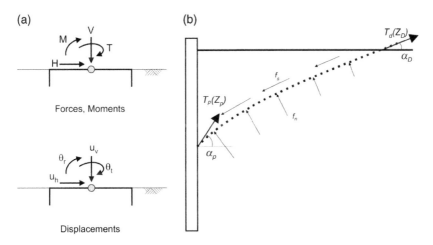

Figure 3.1 Load and displacement sign conventions – positive shown: (a) shallow foundations and (b) anchor mooring.

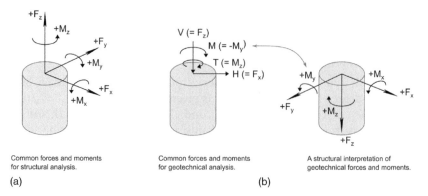

Figure 3.2 Comparison of sign conventions: (a) structural analysis (b) geotechnical analysis – laterally loaded foundations – HM sign convention.

Geotechnical engineers almost universally adopt compression positive with a right-handed axis and clockwise positive sign convention for loads and moments, as shown in Figure 3.1a.

Figure 3.1a shows that an in-plane horizontal load acting left to right and clockwise moment are taken as positive, and this is logical since the overturning moment generally results from an overturning horizontal load acting above foundation level – i.e. both act in the same direction. Some softwares take overturning H and restoring M as positive, so care is needed to change sign convention if this is the case.

Structural and geotechnical engineers can adopt different conventions, and it is important to transfer/convert loads properly. In particular the geotechnical compression positive convention may differ from the structural engineering interpretation that often considers tension positive. This is straightforward in itself, but it also affects the direction of the y axis, which needs to be accommodated in conversion (Figure 3.2). Multiple variations of sign conventions can be adopted by different engineers across the design process, and it is essential that the adopted convention is clearly communicated between groups.

3.3 STRUCTURAL TO GEOTECHNICAL LOAD CONVERSION

Structural engineers generally supply all six structural load components (i.e. F_x, F_y, F_z, M_x, M_y, M_z). Equivalent VHMT geotechnical loads need to be derived from these. As expected, V and T loads are straightforward. However, resultant H and M are slightly more complicated. Using Excel/MathCAD function atan2(x,y), the equations to derive VHMT loads are given by:

$$\alpha_H = \text{atan2}\left(F_x, F_y\right) \tag{3.1}$$

$$\alpha_M = \text{atan2}\left(M_y, -M_x\right) \quad \text{note the minus } sign \tag{3.2}$$

$$V = -F_z \tag{3.3}$$

$$H = \text{sqrt}\left(F_x^2 + F_y^2\right) \tag{3.4}$$

$$M = \text{sqrt}\left(M_x^2 + M_y^2\right); = -M \text{ if } |\alpha_H - \alpha_M| \approx 180° \tag{3.5}$$

$$T = M_z \tag{3.6}$$

where

α_H = resultant H angle x-y plane (radians)

α_M = resultant M angle x-y plane (radians)

Note the use of function atan2 and its arguments. Excel/MathCAD functions atan2(x,y) return the angle (in radians) from the x-axis to a line between the origin (0,0) and the point (x,y). The corresponding Matlab/FORTRAN functions have transposed arguments, namely atan2(y,x). Function atan2(x,y) is superior to function atan(number) which returns angles (in radians) only between −90° and +90° and becomes infinite at ±90°.

3.3.1 Example – load conversion

Convert structural loads $(F_x, F_y, F_z, M_x, M_y, M_z) = (828.79, 128.94, -1012.78,$ $-536.40, 3297.37, -135.07)$ to geotechnical VHMT loads and compute resultant angles α_H, α_M and $|\alpha_H - \alpha_M|$. Units are in [kN, m].

α_H = 8.8 [deg]

α_M = 9.2 [deg]

$|\alpha_H - \alpha_M|$ = 0.4 [deg]

V = −1012.78 [kN]

H = 838.76 [kN]

M = 3340.71 [kNm]

T = −135.07 [kNm]

The next paragraph applies to single isolated foundations only: H and M may be non-co-planar for unsymmetrical co-joined foundations subjected to asymmetric VHM(T) loads.

Good practice (internal, sense check) should compare α_H and α_M values. Because H and M are co-planar, α_H and α_M should be (almost) the same ($|\alpha_H$ $- \alpha_M| \approx 0°$, M and H both overturning or both restoring) or differ by 180°

(M restoring and H overturning or vice versa). In practice, small differences are found (a few degrees). These differences are usually due to either (i) numerical "noise" or (ii) small denominator values in arctangent associated with α values close to either 90° or 270°.

3.4 GEOTECHNICAL STRESSES AND STRAINS

Compressive stresses and strains are generally taken as positive in geotechnical design. This is the other way around from structural analysis, where typically tension is positive.

3.5 COMMENTARY

Always request clarification if unsure about units, especially ambiguous ones like MT (or mT), which may mean metric tonnes and not mega tonnes.

Always state units and sign convention in calculations and reports. Other readers may have different systems.

Chapter 4

Marine geology

4.1 GEOLOGY, SEDIMENT TYPES AND DEPOSITIONAL ENVIRONMENTS

Knowledge of marine geology is important for offshore engineering. Soil type and soil properties are the main parameters for geotechnical engineering, and these are determined by geological processes. Understanding these processes, and the variations of the processes, aids prediction of soil type and behaviour, or what to expect.

Facies are distinctive soil/rock units that form under certain conditions of sedimentation, reflecting a particular process or environment. A 30 m borehole can contain a number of different facies, and their thicknesses can also vary laterally.

Table 4.1 lists the three different marine sediment types. These types are not particularly related to marine environments. Most sedimentation takes place in low-lying areas – usually seas and oceans, but also lakes, rivers, marshes and plains. Marine and terrestrial depositional environments, with typical soils, are compared in Table 4.2. Some soils are more hazardous to installation than others. More marine geology information is given in the textbooks by Seibold and Berger (2010), by Nichols (2009) and in Section 2.3 of Randolph and Gourvenec (2011).

4.2 LATERAL VARIABILITY TOP LAYERS

Both shallow and intermediate foundations are extremely sensitive to the soil conditions found in the top few metres below seafloor, and soil conditions can vary laterally quite rapidly – depending upon the depositional environment. If such conditions are anticipated, then shallow seismic surveys can provide lots of additional information at reasonable cost. Needless to say, at least one seismic line has to intersect a ground truth borehole in order to facilitate correlation/extrapolation.

Table 4.1 Marine Sediment Types (Seibold and Berger, 2010).

Type	Examples	Occurrence
Clastic/ lithogenous	Products of erosion and weathering of ANY kind of rock – sandstone, mudstone, carbonate rock, etc. Silica sand, silts and clays	Fine-grained lithogenous sediments (i.e. clays) are the most abundant of all marine sediments (about 70% by volume)
Biogenic/ biogenous	Products of organisms and their activity Carbonate sand, coral reef breccia, oolites and oozes	Biogenic sediments are widespread and arrive at seafloor, about 30% by volume
Chemical/ hydrogenous	Sediments produced by chemical reactions Halite, anhydrite	Not significant by volume, but widespread

Table 4.2 Depositional Environments.

Environment	Feature	Typical Soils
Marine	River/delta	Fine to medium sand and clay; some organic material
	Tidal flat/lagoon/ barrier system	Organic clay; sand lenses
	Beach/continental shelf	Silica sands
	Deep sea	Normally consolidated clay
Terrestrial	Lakes/swamps	Organic material/peat finely interbedded clays/sands
	Alluvial fans/flood plains	Fine to coarse angular sand and gravel possibly cobbles and boulders some organic material inclined layers
	Dunes (aeolian sedimentation)	Fine well-sorted sand
	Glacial deposits	Overconsolidated (medium dense to very dense) sands and gravels till: directly from ice (firm to hard clay) drop stones
	Outwash plain	Sands and gravels normally consolidated (loose to medium dense)

Because sands are the most challenging to penetrate, and gravels/cobbles/boulders may cause premature refusal, overconsolidated glacial sand sites need thorough investigation in order to adequately define the ground conditions. Prime examples include the North Sea, the Baltic and the US East Coast. Similarly, penetration resistances can exceed the highest expected in locations where the ground, already variable, has been re-worked by ice. Such locations include the Volga outwash deposits in the Northern Caspian Sea, offshore Alaska and the Barents Sea, located off the Norwegian and Russian northern coasts.

4.3 SEAFLOOR CONDITIONS

The following bathymetric information should be acquired:

(i) water depth
(ii) seafloor slopes
(iii) sand waves, dunes or ripples
(iv) scars (icebergs, trawl boards, jack-up footprints, etc.)
(v) current velocity at or near seafloor
(vi) obstructions
(vii) infrastructure (pipelines, well heads, etc.).

Ad (i) Water depth measurements obtained during a normal geotechnical investigation using a combination of drill string, echo-sounder and water pressure sensors are normally for reference purposes only. Hence the water depth should be accurately related to datum for support foundations. For deepwater applications, accuracy considerations are not so critical.

Theoretically, the maximum underpressure must not exceed the water depth +1 atmosphere (100 kPa) in order to prevent cavitation. Hence, in very shallow water (say, less than 10 m water depth), special installation techniques must be considered. Besides limited underpressure, sufficient self-weight penetration (SWP) is needed for a hydraulic seal in order to start the suction assisted stage. Suction foundations have been installed in very limited water depth (0.15 m) in Luce Bay (Houlsby et al., 2006; Villalobos, 2007), in the IJsselmeer, Holland, and in Bremerhaven, Germany. For the Luce Bay project, the foundations (skirt lengths either 1 m or 1.5 m, SWP ≈ 0.1 m) were water-filled, probably causing hydraulic leaks from the interior around the tip to the seafloor outside, and were mitigated by applying underpressure (Villalobos, 2007). Neither water depths nor installation techniques for the last two projects are known. A reasonable minimum SWP value is 0.5 m.

(a)

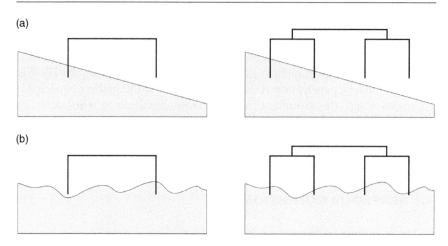

(b)

Figure 4.1 Schematic of (a) sloping seafloor and (b) seafloor ripples.

Ad (ii) excessive seafloor slopes (see sketch in Figure 4.1a), may need levelling.

Ad (iii) for a "sandy" seafloor, seabed ripple heights and lengths should be determined. This is because an essentially level seafloor is required for installation in dense sands – otherwise, local piping may occur if there is limited self-weight penetration, as sketched in Figure 4.1b.

Ad (v) current velocity is required for scour potential in sands.

Ad (vi) the survey should check for the absence of potential obstructions. Examples and more details are given in Section 6.2 (Hazards for Intermediate Foundations).

Chapter 5

Loading conditions and soil drainage

5.1 INTRODUCTION

Effective (not total) stresses affect soil shear strength and volume change characteristics of geomaterials.

It is misleading to think that all "sands" behave "drained" and all "clays" behave "undrained" when loaded. These effects are important. Examples include:

- dilative sands: "undrained" strength can be a factor 3 higher than the "drained" strength
- contractant sand can have little or no strength if rapidly loaded "undrained"
- long-term foundation settlement on clay is a "drained" process.

Soil response depends on:

- soil permeability (coefficient of consolidation)
- soil drainage path length (foundation geometry)
- rate of load application (rapid, slow).

Table 5.1 lists typical soil responses for intermediate suction foundations. It is seen that, for foundation sizing to withstand applied loads, installation and long-term loading conditions generally do not govern. In addition, the soil response is often "undrained" for a range of soil types and load conditions. This is for both "sand" and "clay", and also despite the fact that wave loading has a longer time period (≈ 10 s) than boat impact or seismic loading (< 1 s). Possible exceptions to the "undrained" rule are *italicised* in Table 5.1, and analysis methods for these special cases are given in Section 5.2. The remarks here are solely for "intermediate" caisson foundations: "shallow" foundations, due to their shorter drainage path lengths, and the fact that drainage times are generally proportional to (drainage length)^2 tend to be more "drained" than "undrained".

Table 5.1 Typical Soil Responses for Intermediate Foundations.

Load Type	"Sand"	"Clay"	Remarks
Installation	"drained"	"undrained"	0.001 m/s penetration speed
Wave/anchor cable	*"undrained"*	*"undrained"*	e.g. 100 y extreme storm or 10 y summer storm
Boat impact	"undrained"	"undrained"	–
Seismic	"undrained"	"undrained"	–
Long-term	"drained"	"drained"	consolidation settlement

Note(s): These listings are in order of occurrence, earliest first.
"Sand" could be "partially drained" or "drained" if coarse/gravelly.
"Clay" could be "drained" followed by "undrained" if dead (static) load is significant (e.g. permanent tension).
Section 9.8.4 (CPT Method Coefficient α_u) checks that installation in sand is essentially "drained".

5.2 DRAINED-UNDRAINED

In order to assess whether or not (shallow or intermediate) foundation response should be regarded as "drained" or "undrained", one or more of the following approaches may be considered.

5.2.1 Non-dimensional velocity – penetration rate

A non-dimensional velocity parameter β based on spudcan and CPT tool penetration was proposed by Finnie and Randolph (1994):

$$\beta = v\,D\,/\,c_v \qquad (5.1)$$

where

v = penetration velocity

D = diameter

c_v = soil mass coefficient of consolidation.

Figure 5.1 shows the variation of the non-dimensional bearing modulus with the velocity of penetration, normalised as vD/c_v, where D is the diameter of the foundation and c_v is the coefficient of consolidation, taken as 1×10^{-3} m/s and 5×10^{-5} m/s for the sand and silt respectively. The transition from drained to partially drained occurs at a non-dimensional velocity of about 0.01, while the undrained limit is reached at a non-dimensional velocity of about 30, i.e. if $\beta > 30$, then the soil response is "undrained", and if $\beta < 0.01$, then it is "undrained". The reduction in bearing modulus of the silt

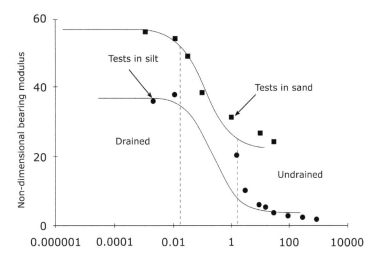

Figure 5.1 Undrained to drained soil response – effect of loading rate on bearing response during spudcan foundation penetration (Finnie and Randolph, 1994).

under undrained conditions is much more dramatic than for the sand, owing to a greater tendency for the latter to dilate at low effective stress levels.

5.2.2 Dynamic drainage factor – dynamic loading of solid piles

Hölscher et al. (2009) and Huy et al. (2009) performed numerical (Biot) analyses and centrifuge model tests of Statnamic Rapid Pile Load Testing in sand and derived a dimensionless dynamic drainage factor η, defined as

$$\eta = GTk/\left(\gamma R^2\right) \qquad (5.2)$$

where

G = soil shear modulus

T = load duration

R = pile radius

γ = soil unit weight

k = soil permeability.

Figure 5.2 presents their results for closed-ended piles. Typical analyses for open-ended pipe piles give η ≈ 1, i.e. a fully drained response (Nguyen et al., 2012).

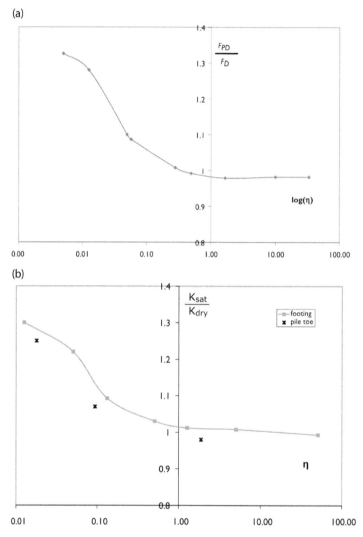

Figure 5.2 Undrained to drained soil response – dynamic loading of closed-ended piles. Centrifuge data (from Hölscher et al., 2009) effect of loading rate on pile tip capacity (their Figure 10) and pile tip stiffness (their Figure 8).

5.2.3 Laterally loaded pile

Osman and Randolph (2012) derived an analytical solution for soil consolidation around a laterally loaded pile, idealised as a circular rigid disk surrounded by elastic soil. Drainage occurs at a finite distance away from the disk. Their equations (verified by 2D plane strain FEA) can be programmed, but are computationally time-consuming. Figure 5.3 summarises their

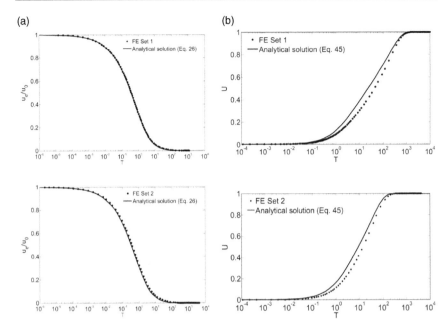

Figure 5.3 Undrained to drained soil response – laterally loaded disk (from Osman and Randolph, 2012). (With permission from ASCE.) Analytical solutions for (a) excess pore pressure dissipation at disk/soil interface T50 ≈ 0.5; T90 ≈ 5 (their Figure 7) and (b) pile consolidation displacement (their Figure 9) T50 ≈ 10; T90 ≈ 75 versus dimensionless time factor $T = c_h t / r_o^2$.

results. It is noted that time factors for pore pressure dissipation are a factor 20 faster than for pile consolidation displacement. Typical analyses for large diameter caissons (D = 10 m) in dense fine sand ("Best Estimate" parameters k = 10^{-4} m/s, E' = 100 MPa, v' = 0.3, c_h ≈ 1 m²/s) give dissipation times t_{50} ≈ 10 s, t_{90} ≈ 100 s, i.e. likely to be "partially drained"–"undrained" for wind or wave loading. For a large diameter monopile foundation (D = 5 m), because dissipation/consolidation times are proportional to D², dissipation times reduce to t_{50} ≈ 2.5 s, t_{90} ≈ 25 s, i.e. "drained"–"partially drained".

Even though c_h can vary by one order of magnitude on either side of the "Best Estimate", these results suggest that laterally loaded intermediate foundations may have to be considered as partially drained when either wind or wave loaded.

5.2.4 Generic

Unlike the Finnie/Randolph and Hölscher/Huy approaches, the Zienkiewicz et al. (1980) method can be used for both static and dynamic problems, and it is not limited to one particular foundation type. One dimensional conditions are assumed – and this may be optimistic. The key diagram is Zienkiewicz et al.'s Figure 3 (not reproduced here). The method first assesses dimensionless parameters π_1 and π_2 using soil permeability k, drainage path length L and loading frequency ω, and then find which Zone (I, II or III) is associated with data point (π_1, π_2). Worked examples include an earth dam response during an earthquake ("undrained" response valid only for k < 10^{-4} m/s at depths well below surface) and pore pressure distribution in seafloor due to wave loading. The Zienkiewicz et al. (1980) method equations can be programmed in MathCAD or similar.

5.3 CLOSURE

No soil in the ground has a membrane around it. Hence partial drainage is always a possibility in all soil types. Even if in-situ tests show drained conditions, due to size effects, the soil may respond undrained or partially drained in the case of large(r) foundations.

Most uncertainty is usually associated with mass c_v (coefficient of consolidation) values to be inserted into analytical equations. In addition, Finnie and Randolph (1994) assume that bearing capacity modulus is proportional to foundation width B. This may not necessarily be the case: other proposals include B^2, e.g. Hölscher et al. (2009), Huy et al. (2009) and Osman and Randolph (2012).

Undrained shear strength may be overestimated when derived from in-situ testing if the soil during testing behaved drained or partially drained. This is particularly the case for CPT q_c using a N_k factor.

If the foundation design is likely to include unconventional soil response, then the geotechnical investigation/laboratory test program should make provision for obtaining reasonably reliable parameter values. Some examples include:

- undrained "sand": permeability k and undrained shear strength s_u (triaxial CU)
- drained "clay": coefficient of consolidation c_v and compression coefficients C_c, C_s and p'_c (oedometer).

If reliable triaxial CU data are not available, then possibilities include: (i) consider a (cautious) drained analysis, (ii) use the DNV (1992) approach using no or limited dilatancy (dilatancy parameter D = 0 corresponds to pure shear in p – q stress space, and D = 1/3 for limited dilatancy) and (iii) use an equivalent undrained shear strength defined by s_u/σ'_{vo} = 0.25–0.35 for a NC clay.

Chapter 6

Hazards, uncertainties and risk minimisation

6.1 INTRODUCTION AND CASE HISTORIES

Intermediate foundations may encounter a range of hazards and are arguably more hazard prone than either shallow or deep (pile) foundations. Examples include:

1) lower-than-expected penetration resistance during installation in NC clay
2) excessive misalignment during installation in NC clay
3) cylinder buckling during installation
4) sand plug liquefaction during installation
5) underpressures close to/above critical during installation in competent sands
6) excessive local scour during operation
7) anchor chain trenching during operation.

Notable case histories illustrating each of these seven hazards are given in the following sections.

6.1.1 Low penetration resistance during installation in NC clay

Two notable cases were first published in 2002 – Girassol (approximately 1400 m water depth, offshore Angola, Gulf of Guinea, West Africa) and Laminaria (water depth ≈ 400 m, Timor Sea).

The Girassol soil conditions are very high plasticity clays ($I_p \approx 110\%$) with organic and carbonate contents around 10%–15% (Colliat and Dendani, 2002). The upper 3 m "crust" is underlain by lightly over-consolidated (LOC) material. The original "Best Estimate" shear strength profile (kPa, m) was $s_{u,DSS} = \max([6 + 1.26 (z - 3)], 6)$ and used for the 3 Riser Tower Anchors (RTA) vertically loaded, and for holding capacity analyses of manifold caissons, FPSO anchors and buoy anchors of various sizes, all VHM loaded.

The 3 Girassol RTAs had been painted externally for protection against corrosion, and their key dimensions were D = 8 m, L = 19.5 m. During the self-weight penetration stage, actual penetrations were higher than predicted and ranged from 15 m to 16 m, compared to 12 m predicted, i.e. only 3.5 m–4.5 m less than the full length (Dendani, 2003). Using over-pressure, two push-out tests performed on one RTA one week and one month post installation gave ≈ 40% and 100% friction increases. These were in relatively good agreement with laboratory thixotropy test data. A revised long-term axial capacity model was developed, assuming "coring", reduced internal friction resistance above internal ring stiffeners, external friction resistance reductions (accounting for the painted steel surface and an upper bound S_t value) and increased thixotropic factor. An additional 175 tonnes of ballast were added to each RTA to obtain the required pull-out capacity.

All Girassol FPSO and buoy anchors had also been painted and moreover had relatively large internal ring stiffeners (0.4 m and 0.2 m outstand respectively). Higher-than-anticipated self-weight penetrations were again measured (Dendani, 2003), and one FPSO anchor was retrieved (i.e. brought back to deck) to assist in contingency decision making. No clay was seen on the painted steel, whereas 20 mm to 30 mm clay thickness covered the non-painted area, confirming that unit friction of painted steel areas should be based on the lower soil-paint (not soil-steel) adhesion. In addition to this, another factor contributing to reduced penetration resistance was the ring stiffener: there was probably a trapped soil wedge below, and a trapped wedge of water above, the stiffener. As with the RTAs, satisfying in-place resistance is more critical than understanding installation. Originally, to meet certifying agency requirements, the FPSO and buoy anchor embedded lengths were designed taking into account tilt and twist (both ±10°), but ignoring reverse end-bearing. Re-analysis, incorporating various combinations of reverse end-bearing (short term), friction "set-up" (longer term) and without tilt (as installed) gave satisfactory anchor holding capacity (Colliat and Dendani, 2002).

The Laminaria soil conditions comprised carbonate silty clay/clayey silt of high plasticity (I_p ≈ 70%) with a carbonate content around 70%. Based on in-situ T-bar and laboratory CAU triaxial compression data, a simplified "Best Estimate" undrained shear strength profile was assessed to be $s_u = 10 + 1.8z$ (Erbrich and Hefer, 2002). To moor the Laminaria FPSO, 3 clusters, each with 3 anchor piles, were used. Pile diameters D were 5.5 m, with embedded lengths L up to 12.7 m, and 20 mm wall thickness (WT). In addition to lug stiffeners, due to the large D/t ratio (275), 8 internal ring stiffeners, each with 165 mm outstand, were used. Their axial spacings varied between 2.25 m and 0.95 m (Figure 6.1). During installation, self-weight penetrations (around 2.5 m) were as predicted. Thereafter, during suction installation, measured underpressures were approximately 50% of

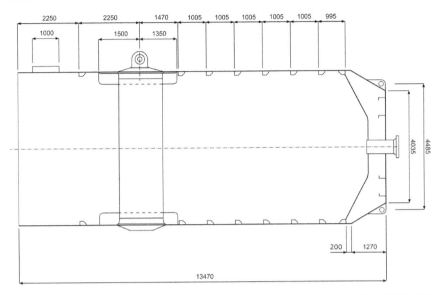

Figure 6.1 Laminara suction pile geometry, D = 5.5 m, L up to 12.7 m, WT = 20 mm, with 8 internal ring stiffeners, outstand = 0.165 m (Erbrich and Hefer, 2002).

the original Lower Bound estimate. Back-analyses suggested that, like Girassol, decreased internal penetration resistance was probably due to a trapped soil wedge below, and a trapped water/soil wedge above each stiffener. Another possibility is a "toothpaste" effect, whereby the inner soil plug is extruded until the height cannot stand unsupported, when it collapses (McNamara, 2000). More recently, Hossain et al. (2012) performed centrifuge testing in OC clay to explore three potential failure mechanisms (Figure 6.2) and provided both a design chart and an equation for back-flow initiation. Anchor in-place holding capacity was unaffected. This was because it was a catenary mooring, θ_{lug} was low, between 15° and 20°, and VH capacity was dominated by lateral resistance.

To avoid on-site embarrassment during installation in weak clays, current design practice is to use reasonable wall thickness to diameter, WT/D, ratios (at least 1/200), and to vary thickness with depth (i.e. highest WT near the top, to accommodate highest underpressures and, if possible, avoid using internal ring stiffeners). In addition, if they are present, internal stiffener outstand widths are kept as small as possible and are also chamfered. Finally, a reduced internal soil-wall adhesion or friction factor, α_i value, is considered above the stiffener to assess (Low, Best and High Estimate) penetration resistances and corresponding underpressures.

Finally, note that both the Girassol and Laminaria projects demonstrate that (in)sufficient V capacity during installation is not necessarily proof of

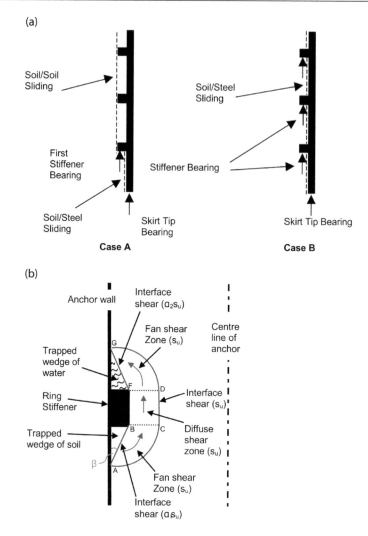

Figure 6.2 Soil flow mechanisms around internal stiffeners during installation of suction caissons (Erbrich and Hefer, 2002).

reduced capacity under VHM loads. This is due to an unaltered lateral resistance H, and load interaction effects; see Chapter 8, Design basis, Section 8.1 (General principles).

6.1.2 Excessive misalignment during installation in NC clay

The Na Kika soil conditions comprise normally consolidated Gulf of Mexico medium plasticity clays. Different undrained shear strength profiles were derived for various design aspects, typically $s_{u,DSS} \approx 1.25z$. To moor the FDS (Floating Distribution) semi-submersible in around 1900 m water depth, a semi-taut mooring system with 4 clusters, each with 4 suction anchor piles, were used. To withstand factored T_{lug} = 10,410 kN at θ_{lug} = 25°, final optimised anchor pile diameters D were 4.3 m, with embedded lengths L = 23.8 m, i.e. a penetration ratio L/D = 5.6, as shown in Figure 6.3a (Newlin, 2003a).

Maximum anchor pile tilt and misalignment values were both ±5°. During installation, self-weight penetrations averaged 13 m, only slightly above the 12.2 m expected. During the subsequent suction assistance stage, all piles remained essentially vertical (tilt < 2°), and 14 of the 16 piles had small misalignments (< 3°). The remaining two piles (L15-P1 and L16-P3) rotated up to 13°, requiring multiple retrieval and re-installation, sometimes with an intentional bias (see Figure 6.3b), before they could be accepted. These two

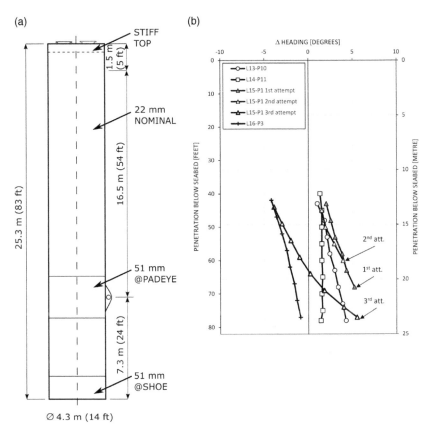

Figure 6.3 Na Kika (a) suction pile anchor sketch (Newlin 2003a), and (b) pile twist versus penetration (from Newlin 2003b). (With permission from ASCE.)

anchor piles were re-positioned at least 26 m away from their initial position, along the anchor radius circumference (Newlin, 2003b).

The large rotations are unexplained and opinions differ: possible reasons include non-vertical lug and internal stiffeners, and currents applying lateral drag to the anchor chain (Lee et al., 2005). Another possibility is unsymmetrical lifting points used for anchor handling. In all cases, at a given penetration, all pile-soil shear has been utilised axially, so a reasonably small torque is needed to cause rotation.

6.1.3 Cylinder buckling during installation

Madsen et al. (2012) describe the 2005 buckling incident in Wihelmshaven, Germany, of an offshore wind turbine (OWT) support foundation in sand, D = 16 m, L = 15 m and WT = 25 mm, i.e. an unusually large D/WT = 640. The installation vessel collided with the foundation at 6.8 m penetration, causing buckling failure. Subsequent installation was not possible, presumably due to a hydraulic leak around the cylinder perimeter between seafloor and the top of the internal soil plug.

Anecdotally, it is understood that at least two anchor pile projects buckled during installation – one in normally consolidated Gulf of Mexico clay (probably due to low D/WT) and the other in the northern North Sea in competent glacial till (probably caused by high underpressures). However, due to legal considerations, nothing has ever been published about their causes, discovery and remedial measures, such that lessons learnt are limited.

6.1.4 Sand plug liquefaction during installation

Senpere and Auvergne (1982) give details of the first ever commercial anchor pile project in the Danish sector of the North Sea in around 40 m water depth. Soil conditions at the 12 locations are sketched in Figure 6.4, and consisted of around 5–7 m of loose to dense sand (φ = 35–40°) underlain by 1–2 m of soft clay (s_u = 20 kPa), and stiff clay (s_u = 70 kPa), from about 6–8 m below seafloor. Suction pile embedded lengths L varied between 8.5 m and 9.0 m, all with D = 3.5 m and WT = 25 mm. The 200-tonne chain design load was applied horizontally at seafloor. CPT q_c values up to 24 MPa in the sand were used to assess penetration resistance. The latter included around 50% reduction due to seepage effects. Because of higher q_c values (up to 30 MPa with 36 MPa peaks), jetting devices were added at tip level just before installation.

During suction assistance, refusal of the first anchor pile occurred at 6.6 m depth due to significant plug heave – the internal soil plug surface had risen by around 3.5 m and had met the top plate. After airlifting the upper sand plug, suction was re-applied and the required 9 m penetration

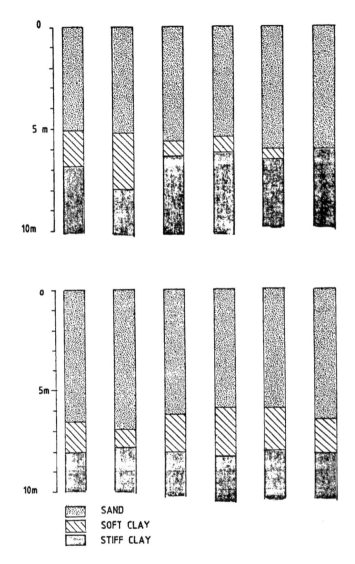

Figure 6.4 Gorm soil profiles. Anchor pile L = 8.5–9 m, D = 3.5 m (Senpere and Auvergne, 1982).

(i.e. tipping out in the 70 kPa stiff clay) was achieved. To install the remaining anchor piles, the top 1.5 m of the soil plug was liquefied by changing the toe jetting system into one below the pile top. No underpressure data were reported.

Based on these data, it is considered that the root cause of pile refusal was due to higher-than-expected penetration resistances, causing high suction pressures and high upwards seepage gradients in the sand plug, leading to liquefaction. Some 35 years later, current practice includes acquiring high quality CPT data well beforehand, using "High Estimate" DNV type k_p and k_f coefficients on the CPT q_c data and to limit underpressures to those which give a seafloor critical hydraulic gradient $i_{crit} = \gamma_{sub}/\gamma_{water}$. At the limit state, a 50% resistance reduction due to seepage is considered not unreasonable (see Section 9.11 on installation in sand).

6.1.5 Underpressures close to/above critical during installation in competent sands

Since Senpere and Auvergne (1982), no cases of liquefying sand plugs have been reported. However, there is a growing body of documentation, both from laboratory permeability tests and from foundation installation, both in the field and centrifuge, of hydraulic gradients $i > i_{crit}$ in medium-dense, dense and very dense sands. Figure 9.15 (in Chapter 9, Installation, retrieval & removal) shows normalised field suction penetration data, plus a theoretical i_{crit} curve. No sand plug liquefaction occurred. Minor heave occurred due to sand plug expansion.

The risk of sand plug liquefaction has been largely mitigated by improved design procedures, sometimes calibrated to measured field data, and adopting due diligence procedures in the field. More details are found in Section 9.11.

6.1.6 Excessive scour during operation

Support foundations in sand likely to be scoured, Figure 6.5 gives an example, are generally protected, usually with some combination of mats, rock fill and sandbags. Occasionally scour protection can fail. A gas platform in around 40 m water depth was founded on four 7 m diameter, 9.9 m long intermediate suction foundations and originally equipped with rock fill placed on mats. The soil profile was sand, the upper 3–4 m of which was loose and mobile. A routine pipeline survey some seven years after platform installation revealed 3 m to 4 m deep scour pits around each suction foundation. Analyses suggested that in-place VHM resistance was insufficient, and the scour pits were infilled with rock fill. This was expensive, since the barge had to be carefully positioned close to a producing platform. Scour protection is discussed further in Section 12.2.

6.1.7 Anchor chain trenching during operation

A more recent hazard is anchor chain trenching in normally consolidated clays. Bhattacharjee et al. (2014) described issues for the Serpentina project offshore Equatorial Guinea. Anchor chain trenches of varying dimensions

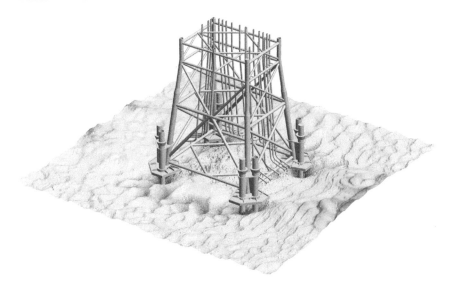

Figure 6.5 Example of local scour resulting from environmental events (copyright Deltares, 2016).

were found in front of all nine suction piles; see Figure 6.6. The piles were considered unfit for service and were replaced by OMNI-Max free-fall anchors.

Detailed 3D FEA of trenches by Alderlieste et al. (2016) have shown that, for a particular Gulf of Guinea project, the anchor pile VH capacity reduction due to trenching was approximately 7% for the design θ_{lug} angles.

More details on this topic, including possible mitigation measures, are to be found in Section 12.12 on anchor chain trenching.

6.2 HAZARDS

Geological hazards applicable to all foundation types have been discussed in Sections 4.2 and 4.3. There are additional potential hazards for intermediate foundations. These may be conveniently subdivided into natural (geological and geomorphological) hazards and those resulting from human activity.

Additional natural geohazards include:

- Cemented layers/rock and coral outcrops. Cemented layers (e.g. beach rock or caprock or just cemented sand). Unexpected cemented layers in sediments may cause refusal (foundation installation) or differential settlement and punch-through risk (foundation performance when loaded).

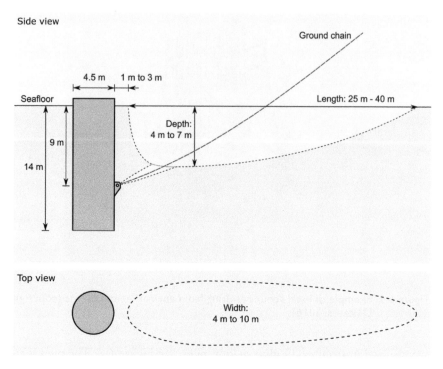

Figure 6.6 Anchor chain trench at Serpentina Field, Gulf of Guinea. Surveyed dimensions (based on Bhattacharjee et al., 2014).

- Undetected boulders in apparently uniform sand (glacial erratics), dropstones in weak clays, coral rubble in carbonate clay, or flintstones in chalk. Besides refusal, such large-sized material may compromise foundation tip integrity (buckling initiation). This is exacerbated since the intermediate suction foundation WT/D (tip wall thickness/ outer diameter) ratio is usually smaller than that for deep (pipe pile) foundations.
- Under- or overconsolidation. Soils with non-hydrostatic pore pressure profiles may consolidate over the foundation lifetime. This is a form of regional settlement.
- Shallow gas. Escaping gas may (a) decrease the resistance of shallow, intermediate and deep (pile) foundations, (b) increase the pressure inside intermediate suction foundations and (c) occasionally cause violent blowouts during drilling conductor wells.
- Sand waves, giving an additional vertical load on the foundation. This is generally unlikely to cause stability or settlement challenges – provided scour protection has been provided.

Additional hazards resulting from human activity:

Human activity at sea has resulted in many obstacles/features on (or under) the seafloor, such as:

- Pipelines, cables, well heads
- Wrecks, archaeological artifacts
- Munitions
- (Jack-up spudcan) footprints
- (Geotechnical) boreholes.

6.3 UNCERTAINTIES

Uncertainties (which are also potential hazards) associated with intermediate foundations can be subdivided into geotechnical data and design categories. They may include but are not limited to the following.

6.3.1 Geotechnical data

The main causes of lack of confidence are usually:

- Inadequate soil investigation. Ideally, any site investigation should be a function of the proposed development/foundation type, and continued until the ground conditions are known and understood well enough for the work to proceed safely (Waltham, 2009). However, intermediate support foundations for oil and gas projects have to be frequently designed using geotechnical data acquired for a piled platform. This is due to changing scales (see Chapter 7, Investigation programs).
- Insufficient laboratory testing, especially for sand
- Unforeseen soil. Examples include:
  - undetected impermeable seams or layers, hindering formation of seepage gradient in sand
  - undetected peat, causing excessive long-term foundation settlements, etc.

If geotechnical data are insufficient, then there are larger than usual uncertainties about:

- soil variability (over support foundation or platform footprint, per suction anchor foundation)
- assessment of soil shear strength
- reliability of soil deformation parameters (for support foundations).

These uncertainties are transmitted into the subsequent design phase, usually as a higher coefficient of variance (COV).

6.3.2 Geotechnical design

Selection of an inappropriate ground model and/or parameter values may result in:

- structural integrity loss (implosion, buckling, handling, etc.)
- insufficient penetration and/or high installation pressures
- excessive self-weight penetration
- excessive tilt and twist
- high (total and differential) settlements.

All of these have been observed in case histories/field experiences, such as illustrated in Section 6.1.

6.4 RISK MINIMISATION

The aforementioned hazards/uncertainties relate mainly to *installation*. Provided the soil type is suitable for installation, insufficient penetration is avoided by the provision of sufficient preload and/or underpressure, combined with a high soil resistance estimate. Using additional steel accommodates high underpressures. Generally excessive self-weight penetration is not necessarily harmful for long-term performance. This is because of VHM interaction effects and/or frictional set-up occurring post installation. However, the installation contractor needs to consider possible effects in their installation methodology. Tilt, twist and settlement are also accommodated during the geotechnical design phase.

The main *in-place* hazards are possible scour and sand waves. Local and general scour both decrease foundation resistance (and stiffness) if no scour protection (rock dump or scour mats) are placed. Also, if the foundation is in abnormal soils (e.g. loose sand which may liquefy during extreme conditions), then risk minimisation options include application of a higher-than-usual Factor of Safety or increasing tip penetration depth in order to encounter more competent soil.

6.4.1 Ground investigation

The principle of any site investigation is that it is continued until the ground conditions are known and understood well enough for geotechnical engineering design to proceed safely (Waltham, 2009).

6.4.2 Geotechnical data

Subject to economic and time constraints, intermediate foundation sites should be investigated as comprehensively as possible. Investigation components include bathymetric, shallow geophysical and geotechnical data, and should be preceded by a desk study. Insufficient attention is often given to desk studies where valuable information can be obtained at low cost.

6.4.3 Laboratory testing

Non-routine onshore laboratory tests on samples for suction foundations normally include:

permeability	sands	suction pump design, in-place resistance
thixotropy	NC clays	retrieval, in-place axial resistance
undrained triaxial/DSS	sands	in-place resistance
cyclic triaxial/DSS	sands and clays	in-place resistance
oedometer	clays	primary consolidation settlement

Section 7.4 (on laboratory testing) provides more details.

6.4.4 Geotechnical design

Judicious selection of soil layering, "Best Estimate" and "High Estimate" parameter values and associated load/material factors account for both installation and in-place uncertainties. Similarly, platform legs may be designed to resist possible jack-up spudcan interaction effects and differential settlements and designing riser–flow line connections to various types of support foundations to resist high total settlement estimates. Large uncertainties generally result in a more robust foundation (i.e. additional steel weight).

6.5 CLOSURE

It has been shown that intermediate foundations are arguably more hazard prone, especially during the installation phase, than either shallow or pile (deep) foundations.

Notable case histories of failure (and their mitigation measures) have been given, the majority of which occurred during installation. Engineers have learnt from their experiences and, due to sharing these, failures have become less frequent since suction foundations were first employed in the early 1980s. Potential hazards for intermediate foundations, both natural and derived from human activity, have also been listed. These may serve as useful checklists during the necessary risk management/risk minimisation process. Appropriate ground investigations and subsequent laboratory testing also assist in reducing project uncertainty.

Chapter 7

Investigation programs

7.1 INTRODUCTION

Any investigation program is a function of the proposed structure – shallow, intermediate or deep foundations, and the anticipated ground conditions. Programs have the following components:

- desk study
- geophysical (seafloor bathymetry, 3D, shallow seismic and side scan sonar)
- geotechnical (and laboratory testing).

7.2 DESK STUDY

Desk studies are often the most cost-effective item in the site investigation process. A desk study may reveal facts that cannot be discovered in any other way. Despite these merits, the desk study phase tends to be skipped too often due to lack of awareness. Good practice is to adjust the scope of work to the intended development, and to bear in mind that more than one foundation type may need consideration. A good example of this is a mooring system, where a variety of anchor types are available (e.g. Senders and Kay, 2002).

Key documents which may serve as useful checklists/templates for geotechnical desk studies are:

- BSI (1999), Clause 6.2 and Annexes A through F
- AGS (2000), Chapter 3
- Simons et al. (2002), Chapter 2
- Randolph and Gourvenec (2011), Section 3.1.1.

7.3 GEOPHYSICAL AND GEOTECHNICAL

General information about planning/requirements can be found in:

- API 2SK (2008), Appendix E.2.1.4, Recommended Sequence for Site Characterization
- API 2GEO (2011), Section 6.2, Shallow Geophysical Investigation
- ISO 19901-4 (2016a), Section 6, Geotechnical Data Acquisition and Identification of Hazards.
- ISO 19901-8 (2014b), Marine Soil Investigations
- ISO 19901-10 (2021) Marine Geophysical Investigations

Comprehensive information about offshore geophysical and geotechnical investigation is to be found in Randolph and Gourvenec (2011), sections 3.2 (geophysical investigation) and 3.3 (geotechnical), and is not reproduced here. Only items relevant to intermediate foundations are discussed in the remainder of this chapter.

Geophysical investigations over large areas are required for both anchor spreads for floating facilities (especially in deep water where the footprint is large, for subsea field architecture and offshore wind farms. They reveal seabed features (e.g. sand waves, potential obstructions) and subsurface ground conditions (e.g. layering, faulting and the presence of shallow gas or hydrates).

Ideally, the geotechnical investigation for an intermediate foundation should be more comprehensive than that for a comparable pipe pile foundation, and comparable in scope to that for a gravity base structure (Gilbert and Murff, 2001). That is, shallow investigation depths, plus at least one cone penetration test CPT per suction foundation (platform or support) or suction anchor (group) location, are necessary. In addition, at least one geotechnical sampling borehole, preferably located at the most heavily loaded leg location (platform), is required to obtain "ground truth" data. Field vane shear tests (with both peak and residual shear strength measurements) are necessary in normally consolidated "clay" profiles. Investigation depths should be at least the estimated penetration depth, plus 1.5 diameters (for large diameter foundations) and to at least around 5 m below estimated penetration depth (suction anchors).

Figure 7.1 illustrates the changing scales between shallow and deep water and also, more importantly, between shallow, intermediate and deep foundations. As shown in Figure 7.1, for a shallow water site, piled platforms are common. Hence the area investigated is concentrated near the platform, and the depth investigated is around 100 m below seafloor. However, for deep water, the scale alters dramatically. Surface vessel anchoring points and seafloor pipelines cover many square kilometres.

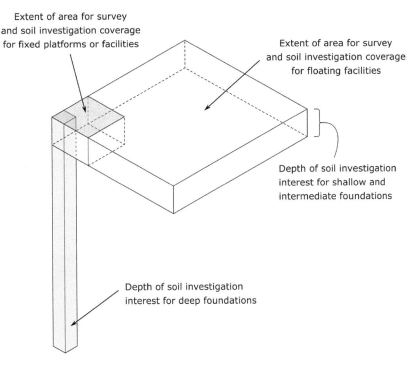

Extent of area for survey
and soil investigation coverage
for fixed platforms or facilities

Extent of area for survey
and soil investigation coverage
for floating facilities

Depth of soil investigation
interest for shallow and
intermediate foundations

Depth of soil investigation
interest for deep foundations

Figure 7.1 Comparative geotechnical survey coverage for shallow or intermediate and deep foundations for floating and fixed structures (after Kolk and Wegerif, 2005).

Moreover, since deep piles are no longer used, the depth investigated is shallow, at most 30 m below seafloor.

Piled foundations are common in shallow water. In this case, geotechnical engineers basically need to know where the competent layer (e.g. dense sand or rock) is to get enough pile axial capacity. Provided that the pile impact driving equipment delivers sufficient energy, installation is not a problem. However, for shallow or intermediate suction installed foundations, geotechnical engineers are interested in relatively subtle variations in the soil layering within the top 30 m. For example, the weaker layer may give in-place resistance problems. Similarly, the harder layer may prevent installation.

In order to minimise risks associated with the hazards listed in Section 6.2, it is occasionally necessary to conduct two separate geophysical surveys. The first is conventional, covering the complete project area. The second is made at the exact foundation locations and is a high-resolution 3D seismic scan to reveal boulders, dropstones, coral rubble, etc.

7.4 LABORATORY

Comprehensive information about the subsequent onshore and offshore laboratory testing programs is again to be found in Randolph and Gourvenec (2011). The relevant sections are 3.4 (soil classification and fabric studies), 3.5 (laboratory element tests) and 3.6 (physical model tests). They are not reproduced here. Several of the more esoteric tests mentioned, such as X-ray examination, palaeontology and geochronology and centrifuge modelling, are not considered for routine foundation design.

Senders and Kay (2002) give requirements for laboratory testing programs specifically for intermediate anchor foundations in deepwater soft clays. A more comprehensive list of laboratory tests, covering soil parameters required for suction caissons in both sand and clays, is in Sturm (2017). The former includes routine classification tests (water content, unit weight, Atterberg Limits, particle size distribution/hydrometer etc.), plus index strength testing (various types of vane tests, pocket penetrometer and fall-cone) are included in the former article, but not in the latter article.

Chapter 8

Design basis

8.1 GENERAL PRINCIPLES

8.1.1 Introduction

Intermediate foundation design usually requires close cooperation between geotechnical and other offshore disciplines, particularly structural engineers.

The following general design considerations indicate how intermediate foundation design can be distinctive from shallow and pile foundation design.

1. Both installation and in-place response need to be considered, usually concurrently.
2. Intermediate foundations must be installed to their design embedment below seafloor. For support foundations, this ensures sufficient in-place capacity to resist combined vertical, horizontal and moment (VHM) loads. Proof of sufficient V capacity during installation is not necessarily proof of capacity under VHM loads. This is due to load interaction effects: V capacity decreases as H and/or M increase. This is unlike conventional piled foundations, where there is almost no interaction between axial and lateral capacity: in such situations, hard driving can usually provide proof of sufficient axial capacity. Similarly, anchors need lug levels at the correct depth below seafloor to optimise pull-out capacity and ensure near-zero rotation. A metre of under-penetration may result in a significant capacity decrease.

   Figure 8.1 gives two commonly encountered examples of installation criticality.
3. An appropriate Factor of Safety on resistance has to be assigned a priori in order to determine foundation geometry. For example, consider the support foundation problem shown in Figure 8.1a. This assumes pure moment (M_0) loading. The second design consideration – that foundations must be installed to their design embedment – uses

factored moment resistance to find the target penetration. Increasing the factor of safety from, say, 1.5 to 2.5 increases M_0 by a factor of 5/3 (= 2.5/1.5) and radius R by 1.186 (= $(5/3)^{(1/3)}$). This results in a significant change in foundation geometry and adds around 20% to the target penetration depth.

Another example is the anchor foundation shown in Figure 8.1b. Consider the case L/D = 5, for which optimum lug level (OLL) is at normalised depth $z_{cl}/L = 0.7$, at which there is no rotation and optimum capacity (100% of maximum) is obtained. If H is applied at, say, $z_{cl}/L = 2/3 = 0.67$, then the decreased capacity is $\approx$ 92% of maximum.

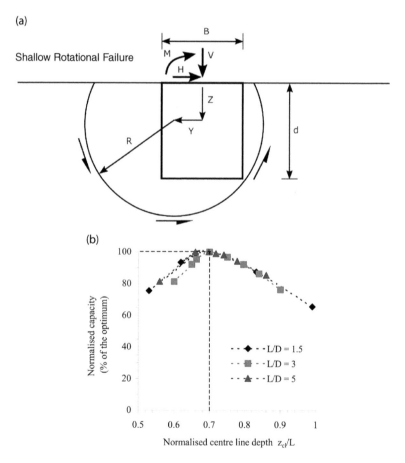

Figure 8.1 Intermediate foundation capacity – sensitivity examples: (a) support foundation – shallow rotational failure, M resistance proportional to R^3 (Kolk et al., 2001); and (b) anchor pile – pull-out capacity decreases if lug not at optimum depth (Supachawarote et al., 2004).

4. Intermediate foundation design usually requires close cooperation between geotechnical and other offshore disciplines, particularly structural engineers.

5. Finally, set-up (axial friction increase) needs to be assessed for taut or semi-taut mooring in normally consolidated clay at relatively short time intervals (say, 10, 30 and 90 days) after installation.

8.1.2 Installation/retrieval/removal

Installation is by dead weight, generally followed by suction assistance. Retrieval and removal are usually by overpressure, possibly with crane assistance. Impact driving, vibratory driving or drilled-and-grouted techniques are sometimes used to achieve high penetrations (L/D up to 10) in competent sand and (weak) rock. Since they are well understood, these techniques are not discussed in detail in this section.

The following general principles need consideration in assessing installation/retrieval/removal of intermediate foundations:

- Driven foundation installation should be analysed using ISO 19901-4 (2016a) Clause 9 (pile installation assessment).
- Suction foundation installation/retrieval/removal can be analysed by limit equilibrium methods ensuring equilibrium between design loads and design resistance.
- Soil resistance models include conventional bearing capacity and CPT cone resistance.
- Penetration mechanism is axial coring.
- Design loads are unfactored.
- Design resistance should be "Best Estimate" (unfactored) and "High Estimate" (factored).

As sketched in Figure 8.2, during installation, due consideration shall be given to the possibility of

- soil plug heave (clay and sand)
- soil plug liquefaction (upwards seepage in sand)
- soil plug reverse end-bearing (clay)
- soil plug separation (clay over sand)
- piping (inwards from seafloor around foundation wall)
- hydraulic fracture (clay over sand).

Where these are critical, more complex analysis approaches are required. Note the following:

- Calculations using alternative analysis methods should include an explanation of any possible differences due to the method adopted.
- Wall and top plate thickness shall be adequate to resist the stresses during installation as well as the axial and lateral in-place loads.

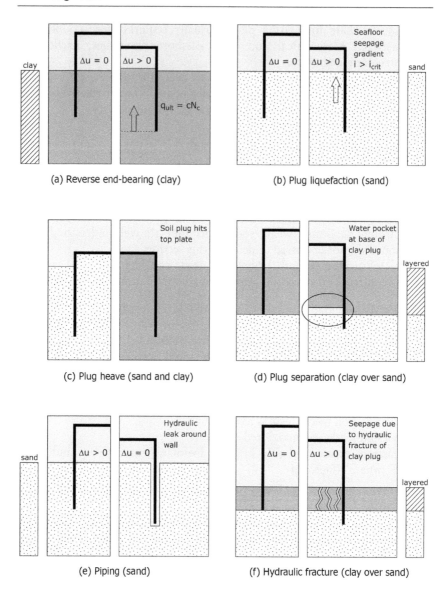

Figure 8.2 Intermediate foundation – geotechnical failure modes during suction installation.

Comparing Figure 8.2c and 8.2d, note that plug heave occurs in both cases and, if large enough, will halt installation. However, the heave is due to the volume of soil displaced by the foundation steel and water pocket formation respectively. Also, water pockets can occur only in clay over sand profiles, whereas soil displacement occurs in all soil profiles.

8.1.3 In-place resistance

Following ISO 19901-4: 2016a Clause 7 for shallow foundations, the general principles given below need to be considered in assessing in-place resistance of intermediate driven and suction-installed foundations:

- Bearing failure constitutes any failure mode that could result in excessive combinations of vertical displacement, lateral displacement, or overturning rotation of the foundation.
- Foundation resistance can be analysed by various methods, including limit equilibrium and yield surface.
- Limit equilibrium methods shall determine the shape and location of the critical failure mechanism. These depend on the design loads, soil stratification and foundation geometry.
- Yield surface methods ensure that factored VHMT loads lie within (or on) a VHMT foundation resistance envelope.
- Due consideration shall be given to the possibilities of excessive displacement and deformation of the foundation soil. Where these are critical, more complex analysis approaches are required.
- Calculations using alternative methods of analysis shall include an explanation of any possible differences due to the method adopted.
- Design loads need to be assessed with due consideration given to the design life of the foundation.
- Seafloor gradient and/or installation tolerance has to be taken into account in design. Tolerable foundation tilt and twist should be specified.
- Undrained calculations shall be adopted where no drainage, and hence no dissipation of excess pore pressures, occurs during loading. This can occur as a result of the rate of loading or the impermeable nature of the soil. In contrast, drained calculations shall be adopted where no excess pore pressures arise during loading. Analysis of foundations subject to partial soil drainage during the loading event is complex, and specialist advice shall be sought in these cases. Impact of structural openings, stiffeners and protuberances shall be taken into account in design.

H and M loads are usually co-planar for single isolated foundations but may be non-co-planar for unsymmetrical co-joined foundations subjected to asymmetric VHM(T) loads (e.g. a braced tripod platform). In the latter case, resultant H and M angles α_H and α_M need to be calculated, for which the equations were given in Section 2.6. If $|\alpha_H - \alpha_M|$, the angle between H and M, is significant, then it is usually cautious (i.e. overestimates in-place resistance) to assume they are co-planar for overturning HM loads, and optimistic if HM loads are restoring. In the latter case, options include applying a higher factor of safety, using 3D FEA on the most critical load cases, or to seek specialist advice. More details of the effect of non-co-planar MH loads are given in the next section.

8.1.4 In-place resistance – non-co-planar MH loads

Figure 8.3a shows schematically two MH resistance ellipse envelopes. Figure 8.3b shows the HM sign convention: as drawn; both H and M are positive and act so as to overturn the foundation. Considering the MH ellipses in Figure 8.3a, it is seen that the effect of non-co-planar MH loads is to increase resistance in quadrant Q1 (HM overturning) and to decrease resistance in quadrant Q4 (HM restoring). The two ellipse shapes suggest that, for the same foundation geometry and soil profile, the MH resistance

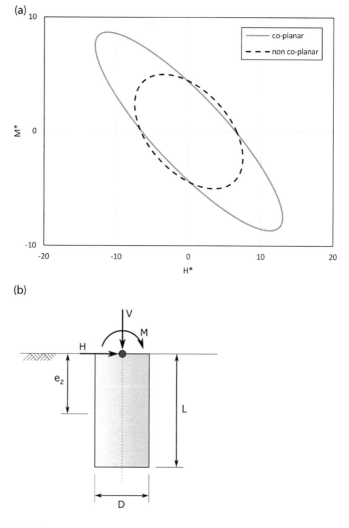

Figure 8.3 MH resistance ellipses for non-co-planar loads.

change is less in Q1 (HM overturning) than in Q4 (HM restoring). Resistance is unchanged when H = 0 and M = 0 (the two MH ellipses intersect on the H and M axes). The next two figures give examples for quadrants Q1 and Q4 respectively.

Figure 8.4 shows a stubby (say, L/D = 1) support foundation subjected to overturning HM loads (i.e. loaded in quadrant Q1). The likely failure mechanism is one of shallow rotational failure, with an approximately spheroidal/circular failure surface passing through the foundation tip. Since the H component is offset from the centreline, non-co-planar HM loading changes the coordinates of the critical (lowest FOS, usually obtained by a grid search) rotation point and increases the failure arc radius. The latter increases the available M resistance. This is for shallow rotational failure, but the same argument can also be applied to the intermediate and deep rotational failure modes shown in Figure 2.2a and 2.2b. Hence non-co-planar HM load increases resistance in quadrant Q1 (HM overturning).

To illustrate quadrant Q4, Figure 8.5 is for a long (L/D = 6, say) caisson anchor subjected to pure HM loading (i.e. V_{load} = 0). Assuming lug level is at optimum depth below seafloor, the lateral load H_{lug} equals H_{max}, the maximum available lateral resistance. Because the foundation is rigid, load H_{lug} is equivalent to loads H = H_{lug} and restoring M = - H_{lug} $z_{lug,opt}$ at

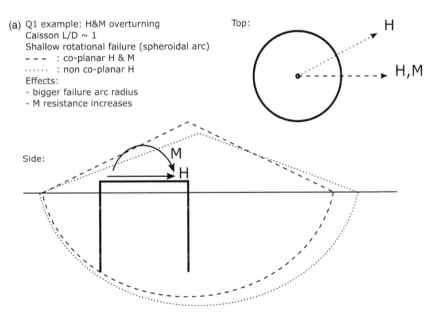

Figure 8.4 Non-co-planar HM loads: (a) Q1 example (overturning HM loads).

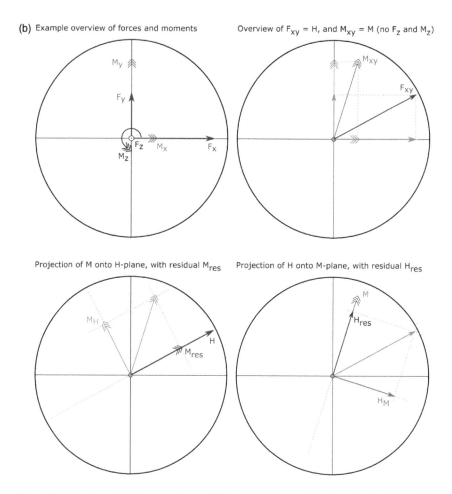

Figure 8.4 Continued: Non-co-planar HM loads: (b) example of projecting H and M.

seafloor. This is the (optimum) situation in Figure 8.3. If a non-co-planar H load is applied at seafloor, then it is obvious that conditions are no longer optimal, and available H resistance decreases. Again, the argument can be extended to combined VHM (instead of HM) loading. Therefore, in quadrant Q4 (HM overturning), non-co-planar HM load decreases available resistance. This situation is not dissimilar to that of anchor pile twist (see Section 10.9), where the maximum available V resistance decreases due to torque T load.

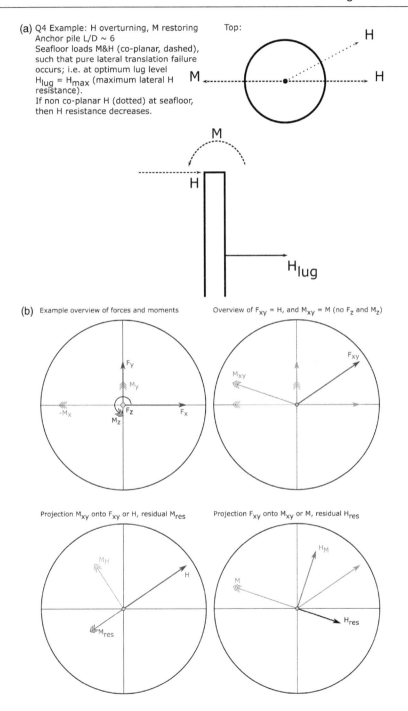

Figure 8.5 Non-co-planar HM loads: (a) Q4 example (restoring HM loads); (b) example of projecting H and M.

8.1.5 In-place response

Following ISO 19901-4 (ISO 2016a), the general principles listed need to be considered in assessing in-place response/serviceability of intermediate foundations:

- Design can be based on in-place response/serviceability (rather than in-place resistance) criteria, whereby the foundation deformation is assessed against allowable movement criteria. The appropriateness of adopting this approach will depend on the type of structure and its installation.
- The selection of appropriate soil moduli (especially considering strain dependency and cyclic loads) is essential in calculation of in-place responses (i.e. serviceability limit states).

8.2 SIGN CONVENTIONS, NOMENCLATURE AND REFERENCE POINT

Vertical (V), horizontal (H), overturning (M) and torsional (T) loads are centric and act at a geotechnical reference point (RP), which is at the mid-point of the foundation at seafloor level; see Figure 8.6. This is the point of

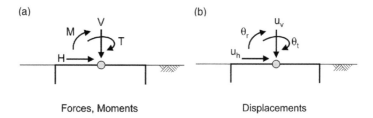

$$\begin{bmatrix} V \\ H \\ M \\ T \end{bmatrix} = \begin{bmatrix} K_{zz} & 0 & 0 & 0 \\ 0 & K_{xx} & K_{x\theta} & 0 \\ 0 & K_{\theta x} & K_{\theta\theta} & 0 \\ 0 & 0 & 0 & K_{\psi\psi} \end{bmatrix} \cdot \begin{bmatrix} \delta_z \\ \delta_x \\ \theta_{xz} \\ \psi_{xy} \end{bmatrix}$$

Figure 8.6 Sign conventions, nomenclature, reference point and seafloor stiffness matrices for analysis of intermediate foundations.

(d)

$$\begin{bmatrix} H \\ M \end{bmatrix} = \begin{bmatrix} K_{xx} & K_{x\theta} \\ K_{\theta x} & K_{\theta\theta} \end{bmatrix} \cdot \begin{bmatrix} \delta_z \\ \delta_{xz} \end{bmatrix}$$

$$K_{HM} = \begin{bmatrix} K_{xx} & K_{x\theta} \\ K_{\theta x} & K_{\theta\theta} \end{bmatrix}$$

Figure 8.6 (Continued)

structural load transfer. Loads H and M are assumed to be co-planar. Associated displacements, vertical (u_V), horizontal (u_H), overturning (θ_M) and torsional (θ_T), refer to the same reference point.

Note that the geotechnical reference point (RP) is the same for shallow and intermediate foundations.

Intermediate suction foundations usually have "stick-up" above seafloor to allow for general seafloor slope, local seabed variations and steel displaced in soil; see Sections 9.10.2 (clay plug heave) and 9.11.1 (sand plug heave). In addition, platform foundations may have platform leg(s) offset from the foundation midpoint, in order to improve barge transport/barge stability. These features mean that specified loads may not act centrically and at seafloor level. Since intermediate foundation geometry (generally outside diameter, stick-up, geotechnical RP and structural RP) varies during design, this challenge cannot be met by a rigid beam element connecting the two RPs. In such cases (where RPs differ), structural loads should be transformed to the geotechnical RP. Similarly, foundation stiffness matrices supplied by the geotechnical engineer should be applied by the structural engineer at the correct offset.

8.3 FOUNDATION STIFFNESS AND FIXITY

Unlike pile foundations, intermediate foundations respond essentially rigidly, i.e. a constant rotation θ_{xz} with depth below seafloor, under lateral HM loads (see Table 2.1 and Figure 8.6b).

For structural analysis purposes, the seafloor foundation stiffness matrix is a simple 4 × 4 symmetrical matrix expressing the relationship between VHMT loads and corresponding displacements δ_z, δ_x and rotations θ_{xz}, Ψ_{xy} as shown in Figure 8.6c. Note the following:

- Both axial V and torsion T loads are assumed uncoupled from the lateral (HM) component. This is unlike in-place capacity (Section 10.10, Resistance under Combined VHM(T) Loads) where there is coupling between V, T and HM. Both V and T resistances are adversely affected by HM loads, V resistance more than T.
- The seafloor foundation stiffness matrix contains coupling between the horizontal and rotational terms. This is because the rotation point is located below seafloor when H and/or M is applied
- The rotation point is assumed to be on the foundation centreline – i.e. small lateral offset is ignored.
- Torsion loads T are usually small and neglected.

Of importance is the 2×2 lateral K_{HM} sub-matrix (Figure 8.6d). Occasionally, some structural jacket analysis programs cannot cope with the off-diagonal terms. In such cases, an equivalent K_{HM} sub-matrix may be redefined using lateral spring K_H and rotational spring K_M, both positioned at the foundation rotation depth h. A rigid beam element connects these springs to seafloor. Terms K_H, K_M and h are simple functions of K_{xx}, $K_{x\theta}$ and $K_{\theta\theta}$ (see Figure 8.7a).

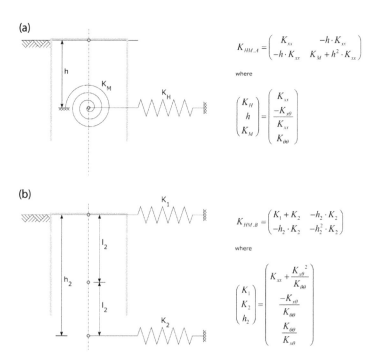

$$K_{HM.A} = \begin{pmatrix} K_{xx} & -h \cdot K_{xx} \\ -h \cdot K_{xx} & K_M + h^2 \cdot K_{xx} \end{pmatrix}$$

where

$$\begin{pmatrix} K_H \\ h \\ K_M \end{pmatrix} = \begin{pmatrix} K_{xx} \\ \dfrac{-K_{x\theta}}{K_{xx}} \\ K_{\theta\theta} \end{pmatrix}$$

$$K_{HM.B} = \begin{pmatrix} K_1 + K_2 & -h_2 \cdot K_2 \\ -h_2 \cdot K_2 & -h_2^2 \cdot K_2 \end{pmatrix}$$

where

$$\begin{pmatrix} K_1 \\ K_2 \\ h_2 \end{pmatrix} = \begin{pmatrix} K_{xx} + \dfrac{K_{x\theta}^2}{K_{\theta\theta}} \\ \dfrac{-K_{x\theta}}{K_{\theta\theta}} \\ \dfrac{K_{\theta\theta}}{K_{x\theta}} \end{pmatrix}$$

Figure 8.7 Seafloor foundation lateral stiffness expressed as (a) horizontal and moment spring stiffnesses K_H and K_M, and (b) non-linear stiffness through springs with K_1 and K_2.

Another situation occurs when lateral spring load-deformation ("p-y") data need modification. Examples include introducing non-linearity (to model varying soil stiffness) or lateral y-shifts (due to spudcan installation). It is easiest to implement these modifications in structural jacket analyses by using a pair of (non-linear) lateral support ("p-y") springs. The first spring has lateral stiffness K_1 at seafloor, the second stiffness K_2 at depth h_2. These two springs are connected by a vertical rigid beam element. Again, K_1, K_2 and h_2 values are simple functions of K_{xx}, $K_{x\theta}$ and $K_{\theta\theta}$ (Figure 8.7b). Lateral stiffnesses K_{xx}, $K_{x\theta}$ and $K_{\theta\theta}$ may be assessed by various methods, including elastic solutions (see Section 11.2.2 on immediate displacement) or from non-linear FEA (e.g. Dekker, 2014).

Finally, more complex studies (e.g. wind turbine monopiles under omni-directional variable cyclic loads) may require the use of a macro element (a non-linear, force-resultant plasticity model, typically a yield surface and back-bone curves accounting for displacement accumulation due to soil stiffness). This is specialised non-routine engineering, generally for detailed case-by-case design using a project specific foundation geometry and soil profile.

8.3.1 Seafloor VHMT loads

Note that:

- axial V load is uncoupled from lateral (HM) load component. In other words, foundation lateral fixity condition does not affect V load.
- lateral H load is statically indeterminate. Hence, the global shear load is equally distributed between all platform legs.
- moment load M is large/important for platform foundations. Note that this is generally overturning for *un-braced* structures (e.g. a self-installing platform or jack-up rig) but may be restoring for *braced* structures (e.g. a conventional platform or a braced met mast or wellhead tripod).
- torsional (T) load is usually small.

8.3.2 Foundation lateral and rotational fixity

As can be seen from the stiffness matrix in Figure 8.6d, lateral *fixity* conditions ($\delta_x \theta_{xz}$) influence lateral HM load magnitudes.

Table 8.1 lists some foundation rotational lateral fixity conditions. When foundations are not "pinned", they attract both horizontal (H) and moment (M) loads. Usually, design practice is as follows: during preliminary design, fully fixed analyses ($k_{xx} = k_{\theta\theta} = \infty$) are performed in order to maximise M load, and then to decrease the stiffness values during final design once a foundation stiffness matrix has become available.

Table 8.1 Foundation Lateral Fixity, Rotation and M loads.

Foundation Fixity Condition	Rotational Stiffness $k_{\theta\theta}$ Rotation θ_{xz}	M Loads
"Pinned"	$k_{\theta\theta} = 0$	$M = 0$
	θ_{xz} = maximum	
"Intermediate"	$0 < k_{\theta\theta} < \infty$	$0 < M <$ maximum
"Fully fixed"	$k_{\theta\theta} = \infty$	M = maximum
	$\theta_{xz} = 0$	

Project experience suggests that, for intermediate foundations, M load is relatively insensitive to rotational stiffness $k_{\theta\theta}$ (and k_{xx}).

8.4 LOAD AND MATERIAL FACTORS

Offshore foundations have been historically subdivided into shallow foundations (Table 8.2) and piles (Table 8.3). Currently there is no industry-wide consensus for (a) definitions of shallow and pile foundations and (b) into which category intermediate foundations appear in Tables 8.2 and 8.3. Hence, factors of safety for support foundations appear in (shallow) Table 8.2, whereas factors for suction anchor piles are taken from (pile foundation) Table 8.3. Note that ISO 19901-7:2013 (ISO, 2014a) lumped FOS values are load angle dependent, more details of which are given in Figure 8.8. Unlike shallow and deep (pile) foundations, there have been no full-scale VHM load tests on intermediate foundations to verify in-place resistance models.

Unlike most USA (API) codes, most European codes differentiate between environmental (live) and static (dead) loading, and also usually specify different load combinations for analysis. This partial factor design (PFD) approach, also known as LRFD – Load and Resistance Factor Design, is more rational than a lumped FOS approach. Figure 8.9 compares both PFD and lumped FOS approaches for support foundations (using a VHM envelope design) and anchor foundations (VH envelope design). In particular, note that where dead load opposes live loads, European code partial load factors may be less than unity. This is useful for anchor foundations, as sketched in Figure 8.9d. Partial load factors per guideline or regulation are included in Tables 8.2 and 8.3. They exclude dead load factors.

8.5 COMMENTARY

Extreme loading conditions relate to probability of occurrence within a fixed time period – e.g. a 10-year summer storm for mudmats/shallow foundations and 100-year (winter) storm for platform foundations.

8.6 CLOSURE

Intermediate foundation design, especially involving suction installation, is very different from that for both shallow and deep (pile) foundations. These major differences have been listed; the most significant of these are possibly: (a) load interaction, since they are subject to high HM loads, and they must reach their design penetration below seafloor; (b) even though installation is usually more critical than in-place resistance, both need to be studied simultaneously; (c) design is multidisciplinary, frequently involving structural engineers to verify steel integrity during installation.

Finally, Figure 8.2 (depicting geotechnical failure modes during suction installation) is a key figure. It neatly illustrates that more things can go wrong during installation than when in-place.

Table 8.2 Partial Load and Soil Resistance Factors for Offshore Shallow Foundations.

Existing Guideline or Regulation (Foundation Type)	Load		Soil Resistance		Overall Factor of Safety[b] (FOS $\approx \gamma_L \cdot \gamma_m$)
	Loading Condition	Recommended Partial Load Factor (γ_L or φ_L)	Recommended Partial Resistance Factor (φ_m)	Recommended Partial Material Factor (γ_m)[a]	
API RP-2GEO (2011) (Shallow foundations)	Not defined				As API RP-2A WSD Section 7.3.1 Uplift: 2.0 Section 7.3.3.2.1 Bearing: 2.0 Sliding: 1.5 Section 6.13.4
API RP-2A, WSD, 21st Ed. (2000) (Shallow Foundations)	Not defined				
DNV (1992) (Jack-ups)	Not defined	1.30 (env.) 1.00 (static) Table 8.1	0.77	Bearing: 1.20 (eff stress) 1.30 (tot stress) Sliding: 1.30 Table 8.1	(env.): Bearing: ≈1.56 (eff stress) 1.69 (tot stress) Sliding: 1.69

ISO 19902:2007(E) 19901-4:2016 (Shallow Foundations)				
Extreme	1.35 (env.)[c] ISO 19902 Section A.9.9.3.1 1.10 (static) ISO 19902 Section 9.10.3.2	0.80	1.25[d] ISO 19901 Section 7.3	1.69 (env.) 1.37 (static)
Operating	1.215 (env.)[e] ISO 19902 Section A.9.9.3.1 1.30 (static) ISO 19902 Section 9.10.3.2			1.52 (env.) 1.62 (static)

Notes:
Refer to actual documents for detailed explanations/definitions/equations.

[a] API applies φ_m to resistance. DNV and ISO apply γ_m to $\tan(\varphi')$ (sands, effective stress) or s_u (clays, total stress). γ_m is equivalent to $1/\varphi_m$ in case of clays, e.g. DNV material factor φ_m of 1.30 is equivalent to API resistance factor γ_m of 0.77.

[b] FOS assumptions: (i) either all environmental or all static load (ii) clay (not sand) profile.

[c] ISO 19902: Section A.9.9.3.1 gives default load factor $\gamma_L = 1.35$ is for GoM. According to Section A.9.9.3.3, higher γ_L values may be in operation elsewhere: e.g. $\gamma_L = 1.59$ (NW Shelf, AUS) and $\gamma_L = 1.40$ (North Sea).

[d] ISO 19901: material factor $\gamma_m = 1.25$ is for combined (VHM) loading. For pure V loading $\gamma_m = 1.5$ (also in Section 7.3).

[e] ISO 19902: $0.9 * \gamma_L$ (default load factor $\gamma_L = 1.35$).

Table 8.3 Partial Load and Soil Resistance Factors for Offshore Piles.

| Existing Guideline or Regulation (Foundation Type) | Loading Condition | Load | Soil Resistance | | Overall Factor of Safety[b] (FOS ≈ $\gamma_L \cdot \gamma_m$) |
		Recommended Partial Load Factor (γ_L or φ_L)	Recommended Partial Resistance Factor (φ_m)	Recommended Partial Material Factor (γ_m)[a]	
API RP-2GEO (2011) (Piles, Section 8)	As API RP-2A WSD				1.5
API RP-2A, WSD, 21st Ed. (2000) (Platform Piles)	Extreme				Section 6.3.4
	Operating				2.0
					Section 6.3.4
API RP-2T (2nd Ed., 1997) (Tension Leg Platforms)	Axial capacity, extreme loading				1.5*B[e]
	Axial capacity, Operating				2.0*B[e]
	Axial capacity, One line damaged				1.5*B[e]
	Lateral capacity				
API RP-2SK (3rd Ed., 2008)[f] (Permanent Mooring Anchors)	Intact	As API RP-2A WSD			Axial 2.0 Lateral 1.6
ISO 19901-7 (2013) Table 7, Section 10.4.3	Damaged				Axial 1.5 Lateral 1.2

			Bearing:	
DNV (1992) (Jack-ups) NPD (1992) (Piles and Anchors)	Extreme	1.30	1.20 (eff. stress) 1.30 (tot stress)	1.69
ISO 19902:2007(E) 19901-4:2016 (Piles)	Extreme	1.35 (env.)[c] 1.10 (static) ISO 19902 Section 9.10.3.2 1.21 (= $0.9*\gamma_{f,E}$) (env.)[d] ISO 19902 Section A.9.3.1	1.25 ISO 19902 Section 17.3.4	1.69 (env.) 1.37 (static)
	Operating	1.30 (static) ISO 19902 Section 9.10.3.2	1.50 ISO 19902 Section 17.3.4	1.81 (env.) 1.95 (static)

Notes:

Refer to actual documents for detailed explanations/definitions/equations

[a] API applies φ_m to resistance. DNV and ISO apply γ_m to $\tan(\varphi')$ (sands, effective stress) or s_u (clays, total stress). γ_m is equivalent to $1/\varphi_m$ in case of clays, e.g. DNV material factor φ_m of 1.30 is equivalent to API resistance factor γ_m of 0.77.

[b] FOS assumptions: (i) either all environmental or all static load (ii) clay (not sand) profile.

[c] ISO 19902: Section A.9.3.1 gives default load factor γ_L = 1.35 is for GoM. According to Section A.9.3.3, higher γ_L values may be in operation elsewhere: e.g. γ_L = 1.59 (NW Shelf, AUS) and γ = 1.40 (North Sea).

[d] ISO 19902: 0.9 * γ_L (default load factor γ_L = 1.35).

[e] API RP-2T: Bias factor B (recommended minimum 1.5) on API RP2A (WSD) FOS (i.e. FOS=2.25 for extreme/one line damaged reduced environmental loading and 3.0 for operating).

[f] API RP-2SK: factor of safety for maximum anchor load determined from dynamic analysis.

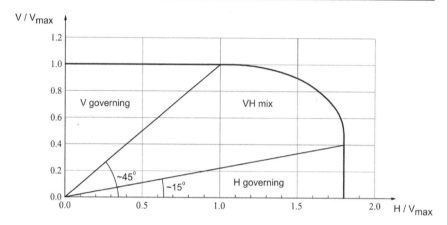

Figure 8.8 Factor of safety versus failure mode for intermediate anchor foundations.

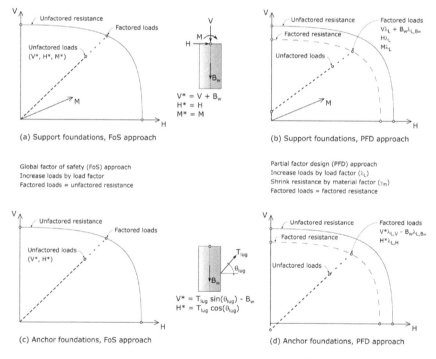

Figure 8.9 Comparison of lumped factor of safety (WSD) and partial factor design (PFD, LRFD) approaches for support and anchor foundations.

Chapter 9

Installation, retrieval and removal

9.1 INTRODUCTION

This chapter is long, emphasising the fact that installation, particularly those methods using suction, can be more challenging than both in-place resistance (Chapter 10) and in-place response (Chapter 11). Topics relating to installation using suction assistance, driven, vibratory and drilled and grouted installation are considered.

Suction installation topics resemble those for shallow foundations, whereas those for intermediate offshore wind turbine monopile foundations are essentially the same as piles.

9.2 GENERAL CONSIDERATIONS – SUCTION ASSISTANCE

Intermediate suction foundations are installed by a combination of self-weight and suction/pumping, and the top plate vent is open. During installation, tilt is small, generally less than 5° at final penetration. In addition, twist is essentially zero. Hence, a simple resistance model is used for penetration, namely an axially coring (unplugged pipe pile) model.

Installation must always be assessed, and is a two-stage process:

- self-weight penetration using foundation dead weight
- suction penetration using pumping.

Retrieval usually also needs to be assessed. In case requirements are not met (e.g. target penetration depth, foundation verticality or orientation), the installation contractor has to use overpressure to extract foundation, bump over (possibly reposition) and reinstall.

Removal is also usually addressed. This is a consequence of site clearance requirements. Removal is performed using overpressure, since this is less expensive than soil removal around the foundation outside perimeter and cutting off the foundation steel, say, 1.5 m depth below seafloor.

Installation feasibility studies include the following general considerations:

- sufficient water depth to create the maximum underpressure
- sufficient steel thickness to resist maximum underpressure (structural)
- additional underpressure costs less than additional preload (steel weight)
- foundations with a high diameter and low embedment ratio are easiest to install using underpressure
- generally feasible in clay, sand and sand over clay profiles
- clay over sand and hard glacial till profiles may be problematic
- cemented sand, rock and "gravel, cobbles and boulders" in sand are generally not feasible.

The conventional procedure for installation analysis assumes static (force) equilibrium, whereby load = resistance. Load is the sum of foundation weight and the product of underpressure and cylinder inner cross-sectional area. The resistance model assumes rigid-plastic ground behaviour and consists of the summation of unit skin friction over the embedded length and unit end-bearing at the toe and internal stiffeners. Internal friction may be substantially reduced above ring stiffeners in lightly to heavily overconsolidated clay if soil is remoulded as it passes over the stiffeners, or if it extrudes, leaving a free-standing column of soil above the stiffener (e.g. McNamara, 2000). In addition, for sand, the effects of water flow are to reduce internal friction and tip resistance, and to slightly increase external friction resistance (e.g. Houlsby and Byrne, 2005). Overall factors to account for flow for installation pressure predictions are also presented in Alderlieste and Van Blaaderen (2015).

Installation assessments assume that both load and resistance act centrally and that the foundation "cores" through the various soil layers during installation, i.e. it remains vertical. However, foundations may possibly tilt (become non-vertical) and/or twist (misalign in plan). Platform-type support foundations, especially if the superstructure is braced, are normally not problematic regarding tilt and twist. This is provided that they are installed on an essentially level seafloor (slope angle less than tilt tolerance) and the suction-assisted penetration depth is sufficient to apply tilt corrections using varying underpressures per foundation (or possibly foundation compartment). However, isolated (single) foundations are more susceptible to tilt and twist – possible causes are items causing eccentric resistance such as lug and chain presence, crane wire torsion and ground inhomogeneity. If tilt criteria are very strict, a possible solution is to install either an instrumented braced foundation cluster (or a single foundation containing compartments), and to vary pressures per foundation (or foundation compartment) in order to minimise/correct tilt during installation. Excessive foundation twist requires retrieval and re-installation (e.g. Newlin, 2003b).

There is no free-fall (uncontrolled penetration) risk during installation. During self-weight penetration, a crane or something similar takes the load, and penetration is usually monitored. During the subsequent suction assistance stage, the suction pump extracts water from the interior at a specified flowrate. Hence, both stages are displacement (not load) controlled. There is no risk of uncontrolled penetration occurring should less competent soil be encountered at foundation tip level during installation. This is unlike pile or jack-up spudcan foundations.

Parameter values for the ground/foundation installation models are usually estimated in two ways – a "Best Estimate" and a "High Estimate". The High Estimate model results are generally used to ensure that the suction foundation is likely to achieve the required penetration depth. Installation parameters are not usually appropriate for in-place analyses. Conventional installation models generally rely upon the results of field measurements, model tests and theoretical analyses. For clay profiles, relevant measurements are those commensurate with penetrating bodies (i.e. CPT) or large displacements (e.g. laboratory remoulded or residual undrained shear strengths, $s_{u,rem}$ or $s_{u,res}$). For sand profiles, the field data base size is small, and actual installation mechanisms are complex and not yet completely understood. Hence, installation predictions are less reliable for sand than for clay profiles. Corresponding removal models have even higher parameter values than for installation. This is due to set-up (outer friction increase) with time: for example, normally consolidated clays regain their original peak undrained shear strength s_u – this can be up to a factor 3 higher (possibly more) than the corresponding $s_{u,rem}$ or $s_{u,res}$ value used for installation.

9.3 GENERAL CONSIDERATIONS FOR MONOPILE INSTALLATION – IMPACT DRIVING, VIBRATORY AND DRILLED AND GROUTED

Most installed monopiles were driven using impact hydraulic hammers (Figure 9.1). This installation method is often the most versatile (can be considered for soft ground to weak rock) and cost-effective solution. At the time of writing, hydraulic hammers with a rated energy going up to 4000 kJ are available on the market. One limitation of this installation method is the noise emission and its impact on marine mammals. Depending on the local regulation, marine mammals protection can require either a very costly noise mitigation system (e.g. bubble curtain), a limitation of the hammer size, a restriction of operation during certain periods or monitoring of marine mammal presence.

Vibratory installation can provide an alternative to reduce the noise emission during monopile installation as vibro hammers are much less noisy, such that costly noise-mitigation measures are not necessary. The industry is gaining more experience with vibro installation, and it is no longer required to complete the last metre of penetration using an impact hammer, as it used to be in the past. As an example, the main objective of the joint industry

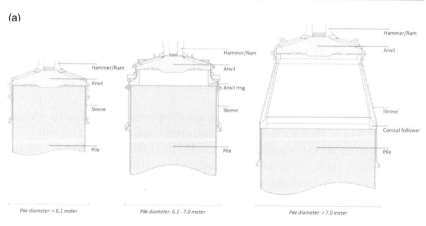

Figure 9.1 Impact driving offshore wind turbine monopiles: (a) cross-over anvils (IHC, 2010), (b) offshore installation examples (Courtesy of Boskalis), (c) hammer as used offshore (Courtesy of Boskalis).

Gentle Driving of Piles (GDP) project is to develop a novel technique for gentle pile driving that simultaneously improves drivability, reduces noise emission and preserves satisfactory geotechnical performance during operations; see also Metrikine et al. (2020). Vibration also tends to create less fatigue in the steel than impact driving. Another advantage of the vibro-installation is the possibility to lift the monopile with the equipment without

Figure 9.2 Drilling from a jackup using RCD (Reverse Circulation Drilling) technique (Courtesy of Fugro).

requiring any external guide (e.g. gripper). Vibro-hammering is, however, not advised in the presence of hard or cemented layers or inclusions and may also not be efficient or possible in certain types of clay.

In the presence of rock layers or thick flint beds, drilling (see the example in Figure 9.2) might be required as a mitigation measure (e.g. in case of potential driving refusal) or as the main installation means. An alternative hybrid installation method of driving and drilling, the so-called Drive-Drill-Drive or 3D method, can be considered to reduce the inner friction or the tip resistance. This can be a cost-effective option by eliminating the need for grout and reducing the size of the hammer spread, and it offers a contingency option for unexpected ground conditions. Under-reaming (i.e. using an enlarged drill-bit to drill to the outside pile diameter) can generally be used to reduce the risk of pile buckling.

However, if the relief drilling operation needs to be performed several times during the installation, the repeated change of tool (i.e. switching from hammer to drilling machine) will considerably increase the installation cost. In addition, due to a large diameter/thickness ratio, monopiles are more susceptible to extrusion buckling than more classical flexible piles. Driving or vibro-hammer may therefore represent a risk in the presence of hard layers, extremely dense sand or very heterogeneous conditions. A drilled and grouted installation can be a safer and more cost-effective method in some cases. Drilled and grouted pile installation involves drilling to the target depth with a diameter larger than the pile, lowering the pile, and grouting the annulus between the pile and the hole. In the presence of overburden layers or upper fractured rock, a support will be required to avoid hole collapse during drilling. Drilled and grouted installation has seldomly been used for monopile installation but could become more common as more complex sites are encountered. Two monopiles of 3.5 m diameters were drilled and grouted on the Blyth offshore windfarm park; these two piles were recently decommissioned. More drilled and grouted monopiles of larger diameter (7 m diameter) should be installed on the French coast in 2022.

Table 9.1 summarises OWT monopile installation feasibility in various soil and rock profiles. More details are given in Section 9.13.

Table 9.1 Comparison OWT Monopile Installation Feasibility in Hard Ground.

Ground profile	Installation method			Remarks
	Impact driving	*Vibratory*	*Drilled & grouted*	*Remarks*
Very dense sand ($D_r \geq$ 85%)	Possible	Possible	Would require support of a casing beforehand to reach bedrock	
Hard clay ($s_u \geq$ 400 kPa)	Possible	To be proven	Possible, but unlikely	ASTM 2487 (2011) /2488 (2009)
Chalk	Possible depending on the chalk grade and presence of thick flint beds	Possible, but unlikely	Possible depending on the chalk grade (see weak rock) or if flint bed are present	CIRIA (2002) C574 classification
Calcareous Sediments	Possible	Impossible	Possible	Kolk (2000)
Weak – MW ($1.25 \leq \sigma_c \leq$ 12.5 MPa)	Feasibility to be checked (depending on other aspects such as rock porosity and fracturation)	Impossible	Possible	ASTM 2487 (2011) /2488 (2009)
MS – ES ($\sigma_c >$ 12.5 MPa)	Impossible	Impossible	Possible	

9.4 BEST AND HIGH ESTIMATES – INSTALLATION RESISTANCE

The additional resistance of internal/external stiffening plates should be taken into account by adding resistance terms accounting for skin friction and end-bearing on each plate.

Installation assessments must use High Estimate (not Best Estimate) soil resistances. Figure 9.3 shows this schematically. The top row, Figure 9.3a, is for in-place resistance assessments (Chapter 10) and plots probability density function (PDF) against (wind, wave and current) loads and (soil) resistances. The PDF for both load and resistance is the same shape, because both have coefficient of variation (COV) values of ≈ 0.3. Because loads and resistances are both factored (see Section 8.4 on load and material factors), the mean values are sufficiently far apart that the probability of failure (i.e. factored loads exceeding factored resistance) is suitably small.

The bottom row, Figure 9.3b and 9.3c, consists of the corresponding PDF graphs for installation (this chapter). Resistance coefficient of variation (COV) is still around 0.3. However, unlike in-place resistance, loads have an extremely low COV, around 0.01. This is because suction pressure and dead weight can be accurately determined. In addition, installation requires failure (loads exceed resistance). Figure 9.3b sketches the situation if the Best Estimate soil resistance and load coincide. It can be inferred that there is a significant refusal risk: for example, at probability $p(f) = 1.00$ on loads, the corresponding probabilities on resistances are < 0.95. However, Figure 9.3c shows that if a High Estimate soil resistance (say $p(f) = 0.95$) is used, the corresponding load probability is 0.5, implying that the risk of refusal has been significantly reduced.

Note that installation assessments should report both High Estimate and Best Estimate soil resistance profiles, together with the corresponding suction pressures. This is to avoid on-site embarrassment when field suction pressures plot below the High Estimate.

9.5 UNDER-PENETRATION AND OVER-PENETRATION

9.5.1 General

This section discusses possible installation measures for both under-penetration and over-penetration of suction foundations.

The objective is to install the intermediate suction foundation (almost) to its design (target) penetration depth (TPD) below seafloor. Organisations have differing approaches regarding target penetration depth tolerance. These account for installation risks such as depth measurement inaccuracy, excessive soil plug heave, encountering obstructions or exceeding the allowable underpressure. Examples include (a) final penetration depth cannot exceed, say, 0.3 m above target penetration depth and (b) defining minimum

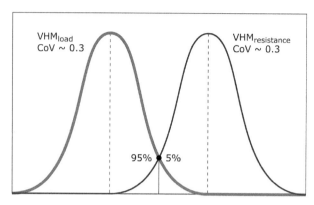

(a) In-place resistance and VHM load

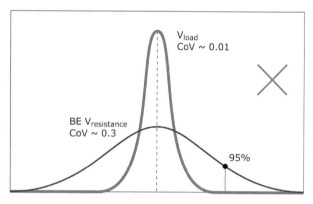

(b) "Best Estimate" penetration resistance and installation load

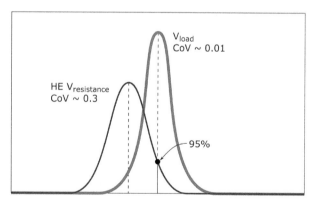

(c) "High Estimate" penetration resistance and installation load

Figure 9.3 Probability density functions comparing loads and (soil) resistances for (a) in-place resistance and installation using (b) Best Estimate and (c) High Estimate soil resistances.

final penetration depth as having ΔFOS (defined as factored resistance/factored loads) value of 1.00, whereas target penetration depth has a higher ΔFOS value (say 1.05). Suction foundations use a combination of self-weight (and possibly preload) and underpressure to achieve target penetration depth.

9.5.2 Under-penetration

Under-penetration is when self-weight and underpressure alone cannot achieve target penetration depth. It usually occurs in competent soils, such as very dense sands or glacial till, where the soil resistance exceeds the High Estimate value. Increasing the underpressure may cause structural damage.

None of the following measures will work if there is a genuine obstruction – e.g. a boulder causing refusal. In such cases, retrieval and re-installation is the only possible measure.

Possible measures to *increase* the final penetration depth include:

- impact driving
- vibration
- jetting (high pressure)
- water injection (low pressure fluidisation)[a]
- air lift
- dredge pump[b]
- explosions
- free-fall (aka drop fall)[c]
- cyclic loading
- coat steel[b]
- friction breakers
- bevel.

Notes:

[a] *water injection*: This fluidises sand at tip level. In clays, it lubricates the clay-steel wall interface.

[b] *dredge pump and coat steel*: Centrifugal dredge pumps can remove the liquefied soil plug contents, and the geotechnical engineer has to design for disturbed soil around the foundation. Similarly, paint, bitumen or other coatings significantly reduce unit skin friction. Both measures are likely to affect axial more than lateral in-place resistance.

[c] *free-fall*: More details, including equations and an example, are given in Section 9.9.3 (on free-fall penetration).

9.5.3 Over-penetration

Over-penetration is when self-weight penetration exceeds target penetration depth, i.e. when pile weight W_{sub} exceeds the Low Estimate soil resistance at TPD. Over-penetration is generally associated with anchor pile foundations in weak clay – the pile weight W_{sub} is heavy and/or the pile-soil α value is lower than the Low Estimate (e.g. the soil is usually more sensitive than anticipated).

The soil resistance increases significantly if the top plate meets the clay soil plug, which is above seafloor due to heave. When this happens, the failure mode changes from "coring" to "plugged". The "plugged" : "coring" installation resistance ratio, η_R, is a function of pile L, D, soil s_u profile and pile-soil α. For a normally consolidated clay, at penetration L, η_R is typically around 2.5 but could vary between 1.5 and 4.5. This resistance increase may be sufficient to halt self-weight penetration (SWP). In this case ($R_{plugged} \geq W_{sub}$) we have L < self-weight penetration < L_{tot}, and the top of the anchor pile remains above seafloor.

If the resistance increase is not enough (i.e. $R_{plugged} < W_{sub}$) then the pile would disappear below seafloor during self-weight penetration, and target penetration depth is exceeded.

Possible measures to *decrease* the final penetration depth during SWP phase include:

- closing the top plate vent(s) just before reaching target penetration depth
- adding an external protuberance (pile outstand/collar) at height above tip equal to target penetration depth
- decrease W_{sub} such that penetration equals target penetration depth, and await set-up.

Ad top plate vent closure: This can be done during offshore installation.

Ad adding external protuberance: This needs to be considered during design. It cannot be done offshore.

Ad decreasing W_{sub}: This may be achieved using crane hook load tension. It is a rather extreme measure – the time required depends on the required frictional resistance increase, and the associated strength gain–time relationship. Alternatively, buoyancy or flotation blocks can be attached to the foundation to reduce the submerged weight.

9.6 DIFFICULT SOIL PROFILES

The following paragraphs give more details of soil profiles likely to be difficult/problematic using underpressure. They are listed in easiest (feasible) to hardest (impossible) order.

Loose sand – easy to install but in-place resistance (liquefaction) considerations

Clay, sand and sand over clay – generally feasible

Calcareous/carbonate clays require no special considerations. This is unlike carbonate sands (and silts), which are usually cemented and weaker than silica/quartz sands.

Sand with occasional clay seams/layers and *sand with occasional peat seams/layers* – Thin clay (or peat) seams or layers, if sufficiently thick and present over the whole foundation footprint, may cause a flow seal and prevent inner friction reduction. Hence, installation may be problematic. At the time of writing there is no consensus as to what clay thickness actually causes no flow – estimates vary between 0.01 m and 0.1 m. Complicating factors are (a) a CPT tool cannot detect clay seams, detection limit is around 2 to 4 tool diameters (b) whether or not the clay is continuous across the majority of the foundation footprint – the thicker it is, the more likely it is not a lens and increases the probability of this happening (c) whether or not the clay becomes punctured by the foundation, causing a hydraulic leak (inducing fluid flow) at the soil/foundation interface. An additional consideration for peat is long-term creep settlement.

Hard glacial till – these are clays containing sand and gravel. Installation may be problematic/impossible if there are frequent cobbles or boulders – foundation tip integrity is likely to be compromised. Design challenges are usually a high underpressure during installation, a domed (not flat) top, and a relatively high wall thickness. An example is the Skarv Field suction anchors (Langford et al., 2012). Soils were soft marine clays underlain by hard glacial till (s_u up to 340 kPa). The caissons were 5 m diameter, up to 14.5 m high and about 40–50 mm wall thickness. The domed tops were designed to withstand the maximum client specified value of 1500 kPa underpressure. Measured values were up to 1000 kPa during installation.

Carbonate sands – these contain > 90% calcium carbonate ($CaCO3$). They are usually cemented and of variable cementation. They are also weaker than silica/quartz sands. Note that cementation may have been destroyed by borehole drilling/sampling. Hence the main risk is refusal (or tilt) during installation, if highly cemented sand is unexpectedly met under (part of) the foundation tip. In addition, in-place axial skin friction is low.

Clay over sand – installation may be problematic because, when the foundation tip enters the underlying sand, there may be no water flow to reduce friction and tip resistance. In such cases, key design challenges are limited penetration, higher underpressures and a higher-than-usual wall thickness. Due to limited penetration, in-place resistance may be insufficient. This is discussed further in Section 9.12 (on installation in mixed sand and clay).

Cemented sand, rock, cobbles and boulders in sand – installation is generally not feasible, as resistance is too high and there is insufficient self-weight penetration. Use other installation methods – e.g. impact driving. A common example is caprock in the Middle East.

Peat – usually of limited thickness and encountered above target penetration depth. Main design challenges include:

- characterisation (e.g. difficult to accurately determine peat strength)
- likely to seal fluid flow (if sandwiched between sand layers)
- (very) low specific weight (risk for the suction process, both seal and installation)
- long-term response (creep settlement component)
- possible low pH value (additional cathodes may be necessary).

9.7 MAXIMUM PUMP UNDERPRESSURE

Houlsby and Byrne (2005) have listed the following practical limits to the maximum attainable suction pump underpressure:

- the absolute pressure at which the water cavitates (usually a small fraction of atmospheric pressure)
- the minimum absolute pressure that the given pump design can achieve
- the minimum relative pressure that the pump can achieve.

In shallow water depth (less than ≈ 30 m), taking allowable underpressure inside the foundation as atmospheric or vacuum can make an appreciable difference to installation in difficult soil profiles, especially competent ones. Installation feasibility may hinge on maximum underpressure – and whether or not the suction pump can cope with possible cavitation (see Section 4.3).

Hence, unless there is convincing evidence to suggest otherwise, it is cautious to take atmospheric (rather than vacuum) pressure.

Water depth is usually related to Lowest Astronomical Tide (LAT).

9.8 PENETRATION RESISTANCE ASSESSMENT

9.8.1 Introduction

There are two main methods of assessing penetration resistance R:

$$\text{CPT } q_c \text{ based} \qquad \begin{aligned} f &= k_f\, q_c\,(\text{sand and clay}), \\ q &= k_p\, q_c\,(\text{sand and clay}) \end{aligned} \tag{9.1}$$

$$\begin{aligned} &\text{classical bearing capacity}\,(\text{alpha / beta}) \\ f &= \alpha s_u\,(\text{clay}),\ f = \beta\sigma'_{vo}\,(\text{sand}), \\ q &= N_c s_u\,(\text{clay}),\ q = N_q\sigma'_{vo}\,(\text{sand}) \end{aligned} \tag{9.2}$$

Of these, the CPT method is considered to be the more reliable/popular. This is especially true for sands, where resistance assessments are critical. This is because the CPT method directly correlates penetration resistance R to cone resistance q_c, whereas two steps are needed for classical bearing capacity, firstly assessing φ' from relative density D_r (or q_c) test data, and then selecting an appropriate N_q value; in this process N_q increases exponentially with φ'. However, the classical bearing capacity method can serve as a useful double check for more challenging sites; for example, those where friction reduction above internal stiffeners has to be considered. A disadvantage of the CPT method is that two sets of parameter values are needed (q_c, etc. for installation; s_u for in-place resistance) whereas classical bearing capacity uses s_u for both installation and resistance, and it is easier to match installation resistances after 30 or 90 days with in-place axial resistance assessments.

Penetration is a "coring" mode (Figure 10.1). Hence, axial soil resistance is the sum of end-bearing on the caisson tip, plus inner and outer skin friction. Other major assumptions (which are conventional) include (a) static equilibrium and (b) "friction fatigue" (unit skin friction decrease with increasing penetration) is not considered, and (c) unpainted steel (d) CPT method assumes adhesion is the same in self-weight penetration and suction assistance zones but classical bearing capacity makes a distinction.

The area over which underpressure/overpressure Δu acts is A_i, the internal area (not the caisson base area $A = \pi D^2/4$). For no water flow, at depths below self-weight penetration, Δu is dependent on W_{steel}, the foundation submerged weight, and Δu values are computed assuming static force equilibrium (load = resistance) from the simple equations:

$$\Delta u = \left(R - W_{steel} \right)/A_i \quad \text{(installation)} \tag{9.3}$$

$$\Delta u = \left(R + W_{steel} \right)/A_i \quad \text{(exaction, i.e. retrieval and removal)} \tag{9.4}$$

where

R = penetration (or extraction) resistance

W_{steel} = caisson submerged weight (and preload, if any)

A_i = caisson inner area = $\pi D_i^2/4$

D_i = caisson inner diameter

Fortunately, since the majority of soil profiles are clay, there is no water flow, and the resistance versus depth profile is usually simple to calculate. Entering the penetration resistance curve with W_{steel} gives the self-weight penetration. At greater depths, the suction pressure is computed directly from the resistance curve using equilibrium Equation (9.3).

However, if either underpressure or overpressure create fluid flow, then penetration resistance R is a function of Δu. An iterative technique is required since Δu appears on both sides of both load–resistance equilibrium equations.

Values of Δu need to be compared with limiting underpressures for base failure (clay) and liquefaction (sand), details of which are given in Sections 9.10 and 9.11. In addition, for shallow water depths, the maximum attainable underpressure (Section 9.7) also has to be checked.

9.8.2 CPT q_c method

CPT cone resistance q_c data are frequently used to predict skirt penetration resistance. This is reasonable since the prototype skirt and CPT are of similar size (as is the case here). Foundation skirt tips are usually flat, whereas the CPT tool is conical. Hence coefficient k_p, relating unit end-bearing q_w to CPT q_c is a shape factor, and has a value of less than 1.

Penetration resistances R are assessed based on the DNV (1992) method. For a uniform wall thickness (WT) cylindrical foundation, R is calculated using the equation

$$R = A_p k_p q_{c,L} + A_{si} \int k_f q_{c,z} dz + A_{so} \int k_f q_{c,z} dz \qquad (9.5)$$

where

L = foundation tip depth

k_f = empirical coefficient relating q_c (or q_t) to unit skin friction resistance

k_p = empirical coefficient relating q_c (or q_t) to unit end-bearing resistance

A_p = foundation wall tip area = $\pi (D^2 - D_i^2)/4$

A_{si} = caisson inner perimeter = πD_i

A_{so} = caisson outer perimeter = πD

D = foundation outer diameter

D_i = foundation inner diameter = D – 2 WT

$q_{c,z}$ = CPT cone tip resistance at depth z below seafloor

$q_{c,L}$ = CPT cone tip resistance at caisson embedded/tip depth L

In Equation (9.1), inner and outer skin friction components have been separated because of differing fluid flow effects. Both terms are integrated from seafloor to depth L. The DNV (1992) method uses *average* CPT cone resistance q_c profiles, plus Best Estimate and High Estimate empirical coefficients k_p and k_f relating q_c to unit end-bearing and skin friction, to derive foundation penetration resistance. DNV RP-H103 (2014) repeats the DNV (1992) coefficients, but a complicating factor is that assessment is done on a *single* CPT. The DNV (1992) averaging approach is preferred – CPTs in sand can vary significantly over short distances – provided that sufficient CPTs have been made.

For sand, DNV (1992) indicates that experience has shown that coefficients k_p and k_f for the upper 1 m to 1.5 m should be 25% to 50% lower than those given in DNV (1992) where local "piping" or lateral platform movement has occurred.

9.8.3 CPT method coefficients k_p and k_f

Based on Equation (9.1), "most probable" and "highest expected" penetration resistances can be calculated. Modern-day practice is gradually progressing towards probability-based assessments. Hence, Best Estimate and High Estimate have generally replaced the original DNV (1992) "most probable" and "highest expected" terms. In addition, the latter DNV term misleadingly implied that the value will not be exceeded.

Table 9.2 Values of kp and kf Coefficients for Sand and Clay.

Soil type	"Best Estimate" R_{BE}		"High Estimate" R_{HE}		Remarks
	k_p	k_f	k_p	k_f	
Clay	0.4	0.03	0.6	0.05	DNV (1992) for
Sand	0.3	0.001	0.6	0.003	"North Sea Conditions"
Clay	0.4[1]	0.006	0.6[1]	0.018	Colliard and Wallerand (2008) for GoG NC clay, f = k_f $q_{T,CPT}$
Carbonate Silt	Most accurate fits from SWP back-analysis Unit end-bearing: either $q_w = 0.6$ $q_{T,CPT}$ or $q_w = 1.0$ $q_{T,T-bar}$ Unit skin friction: $f_i = 1$ kPa[2], $f_o = s_{ur,LE}$ ≈ 1 kPa				Frankenmolen et al. (2017) for Australian North West Shelf

Notation:
unit end-bearing resistance $q_w = k_p$ q_c
unit skin friction resistance $f = k_f$ q_c or k_f q_T
$q_{T,CPT}$ = total cone resistance[3]
$q_{T,T-bar}$= total T-bar resistance
$s_{ur,LE}$ = low estimate soil remoulded undrained shear strength

Notes:
[1] from DNV (1992) – k_p values not stated by Colliard and Wallerand (2008)
[2] low internal f due to tip stiffener
[3] $q_{T,CPT} \approx 1.25$ q_c for NC clay

Table 9.2 gives k_p and k_f coefficient values for various soil types. For North Sea conditions (presumably competent dense sands and stiff clays), coefficients k_p and k_f were first proposed by DNV (1992). Lower k_f coefficients were suggested by Colliard and Wallerand (2008) for Gulf of Guinea clay (normally consolidated, sensitive, high plasticity). Since penetration in normally consolidated clay is governed by skin friction, the Gulf of Guinea clay k_p values are considered to be the same as for North Sea conditions. Germanischer Lloyd (2005, 2013) requires higher High Estimate coefficients than DNV (1992). More recently, Frankenmolen et al. (2017) back-analysed the Prelude project caisson self-weight penetration data, and their most accurate fit values (neither Best Estimates nor High Estimates) are also given. Carbonate soils are notoriously difficult to characterise: penetration resistances were only slightly higher than the Low Estimate but a factor ≈ 5 lower than the High Estimate. Note that the presence of any cemented seams within the silt would have prevented suction installation. Finally, Table 9.2 shows that there is room for improvement – especially in sand, where end-bearing is generally the larger resistance component, and the Best Estimate and High Estimate k_p values vary by a factor 2.

Table 9.3 compares North Sea, Gulf of Guinea and South China Sea clay characteristics. This may be useful when considering extrapolating Colliard and Wallerand (2008) k_p and k_f coefficients to other sensitive, normally

Table 9.3 Comparison North Sea, Gulf of Guinea and South China Sea Clays.

Soil parameter	Units	North Sea clays Lunne and St John (1992)[1]	Gulf of Guinea clays Puech et al. (2005)	South China Sea clays Palix et al. (2013)
General		Competent over consolidated low-medium plasticity insensitive	Weak normally consolidated high plasticity sensitive	Weak normally consolidated high plasticity sensitive
q_c	[MPa]	≈ 2	≈ 0.015 z to 0.030 z [q_{net} gradient]	≈ 0.017 z to 0.020 z [q_{net} gradient]
q_T	[MPa]	n/a	n/a	n/a
q_T/q_c	[–]	≈ 1.1	≈ 1.25	n/a
R_f	[%]	n/a	≈ 0.3 to 0.4[2]	≈ 2 to 3
B_q	[–]	n/a	≈ 0.5	n/a
w	[%]	≈ 15 to 25	≈ 150	≈ 150
w_p	[%]	≈ 15 to 25	≈ 40 to 50	≈ 40
w_L	[%]	≈ 40 to 65	≈ 120 to 160	≈ 80
I_p ($= w_L - w_p$)	[%]	≈ 25 to 30	≈ 70 to 120	≈ 45 to 55
γ	[kN/m³]	≈ 20	≈ 13 to 14.5	≈ 13 to 14.5
s_u	[kPa]	≈ 150 to 400 [very stiff to hard]	≈ 1.5 z	≈ 1.3 z
S_t	[–]	≈ 1.2 to 1.5	≈ 2 to 5	n/a (T-bar unreliable, S_t ≈ 8.5 at 25 cycles)
s_u/σ'_{vo}	[–]	≈ 4 to 10	≈ 0.4 to 0.5	≈ 0.35 to 0.3 ($s_{u,DSS}/\sigma'_{vc}$)

Notation:
q_c = cone resistance
q_T = total cone resistance
R_f = friction ratio
B_q = pore pressure ratio
w_p = plastic limit
w_L = liquid limit
I_p = plasticity index
γ = unit weight
s_u = undrained shear strength
S_t = sensitivity
n/a = not available
z = depth below seafloor

Notes:
[1] Figures 2, 4 and 5 (CPT q_c) and Table 1
[2] Probably a typographical error: R_f is typically 3% to 4% in (sensitive) clays

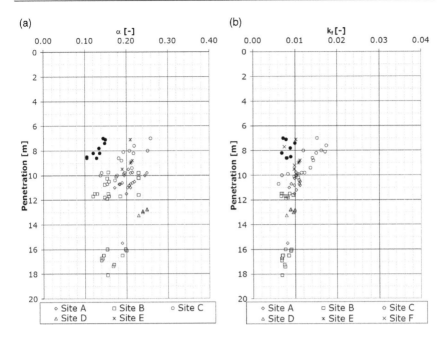

Figure 9.4 Self-weight penetration back analyses in NC Gulf of Guinea clay: (a) adhesion factor α and (b) k_f coefficient (Colliard and Wallerand, 2008) © IHS Markit.

consolidated clay locations (Figure 9.4). The two normally consolidated clay sites are broadly similar, but the devil is in the detail: the open structure of the South China Sea clay gives a higher sensitivity S_t. Low thixotropy and rate effects were also noted. Unlike the Gulf of Guinea, these have possible implications for soil cyclic shear strength degradation, and anchor capacity increases with time. They also underline the necessity of obtaining appropriate ground truth data for use in detailed design.

For soils intermediate between clay and sand (i.e. non-carbonate silt), if uncertainty exists about drainage, then it is reasonable to take the higher of clay and sand coefficients – i.e. use sand k_p for end-bearing and clay k_f for skin friction. Similarly, k_p and k_f coefficients intermediate between the tabulated values for clay and sand should be used for highly stratified soils, interbedded clays and sands, or sand/clay mixtures.

9.8.4 CPT method coefficient α_u

For analyses involving water flow in permeable soil (sand), values of foundation tip pore pressure coefficient α_u are required. These values depend on foundation penetration, the soil plug : soil mass permeability ratio and soil layering. Industry practice assumes steady state and isotropic permeability. Both assumptions are considered very reasonable: steady state is checked at the end of this section, and Dutch onshore practice (mainly excavation dewatering) assumes $k_h = k_v$ in their Pleistocene sands. The results presented in this section need modification if the foundation is not axisymmetric. Limiting α_u values are 0.5 when penetration ratio L/D is almost zero and 0 when the foundation tip enters an impermeable layer.

As shown in Figure 9.5a, Houlsby and Byrne (2005) used finite element analyses for foundations with penetration ratio L/D values up to 0.8 to derive the following approximate equation for isotropic permeability:

$$\alpha_u = c_0 - c_1 \left[1 - \exp\left(-(L/D)/c_2\right) \right] \tag{9.6}$$

where
$c_0 = 0.45$, $c_1 = 0.36$ and $c_2 = 0.48$

Sand plug heave (i.e. loosening) increases the plug permeability. Assuming head loss is reduced in inverse proportion to k, and correcting a typographical error in the denominator, the corresponding Houlsby and Byrne equation is:

$$\alpha_u = \alpha_1 k_f / \left[(1 - \alpha_1) + \alpha_1 k_f \right] \tag{9.7}$$

where
$\alpha_1 = c_0 - c_1 [1 - \exp(- (L/D) /c_2)]$
k_f = sand plug : soil mass permeability ratio [–]

Results for $k_f = 2$ and 5 are also given in Figure 9.5a. Using Equation (9.7), extrapolated α_u values are 0.14 and 0.32 for $k_f = 1$ and 3 respectively at L/D = 1.

Figure 9.6 shows four steady-state flownets obtained for a caisson foundation with L/D = 1, showing equipotentials at intervals of 10% of the change in pore pressure. A slightly modified version of the finite element program P72 (Smith and Griffiths, 1998) was used, which solves Laplace's equation over an axisymmetric region. A 15-by-15 mesh of 4-noded rectangular quadrilateral elements is used, with no element integration and assembly for the cut-off (foundation wall) elements.

The first flownet, Figure 9.6a, is for an infinite half space of uniform permeability. Note that most of the pressure drop occurs within the soil plug between tip level and seafloor. In addition, the hydraulic gradient is reasonably uniform both radially and axially within the soil plug but is a maximum at seafloor. At tip level, $\alpha_u \approx 0.14$ (taken by eye between the equipotential at 0.1 and 0.2), i.e. agrees with Equation (9.7) $\alpha_u = 0.14$.

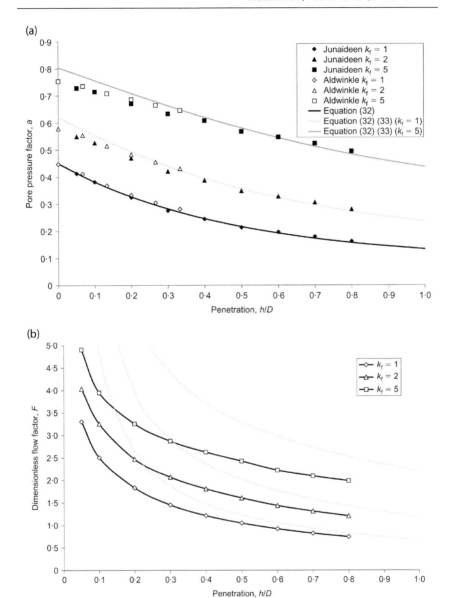

Figure 9.5 Installation in sand with seepage flow: (a) foundation tip pore pressure coefficient α_u versus penetration ratio L/D and (b) dimensionless flow parameter F versus L/D. Flow rate into sand plug Q = F (kDΔu/γ_{sub}). Parameter k_f = soil plug: soil mass permeability ratio. Use the bottom-most curve (k_f = 1) for uniform permeability (Houlsby and Byrne, 2005).

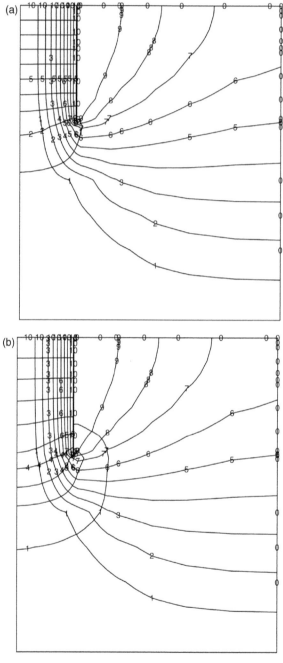

Figure 9.6 Flownets due to underpressure/overpressure. Axisymmetric steady state fluid flow. Foundation penetration ratio L/D = 1.0, tip pore pressure coefficient α_u and dimensionless flow parameter F: (a) uniform permeability k, infinite half space, $\alpha_u \approx 0.14$, F $\approx$ 0.60. (b) plug k : mass k = 3, infinite half space, $\alpha_u \approx 0.30$, F $\approx$ 1.30.

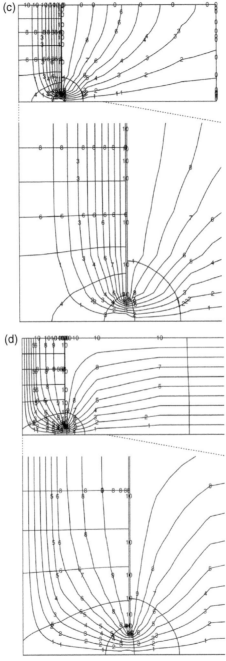

Figure 9.6 Continued: Flownets due to underpressure/overpressure. Axisymmetric steady state fluid flow. Foundation penetration ratio L/D = 1.0, tip pore pressure coefficient α_u and dimensionless flow parameter F: (c) uniform permeability k, impermeable boundary at 1.1 D bsf. $\alpha_u \approx 0.24$, F ≈ 0.50. (d) confined aquifer: as (c) but with impermeable suface. $\alpha_u \approx 0.32$, F ≈ 0.42.

In the real world, the soil is usually of non-uniform permeability and the far field boundaries are not at infinity, and sensitivity analyses of such routinely encountered cases are given on the remaining three components of Figure 9.6. Figure 9.6b shows the flownet for the case when $k_f = 3$ (was $k_f = 1$). Due to the reduced head loss in the sand plug, the tip pore pressure coefficient α_u increases, $\alpha_u \approx 0.30$ (was $\alpha_u \approx 0.14$). This means that the plug seepage gradient reduces (for the same L/D and underpressure Δu). Again, $\alpha_u \approx 0.30$ agrees well with Equation (9.7) $\alpha_u = 0.32$.

Figure 9.6c explores the case when the foundation tip approaches an impermeable boundary. It shows the flownet when the impermeable base is no longer at infinity, but at 1.1D below seafloor (bsf) (i.e. 0.1D below tip level). As expected, reduced radial transmissivity below the tip increases the head drop in this zone, again leading to an α_u increase (now $\alpha_u \approx 0.24$, was $\alpha_u \approx 0.14$).

Finally, Figure 9.6d has the same boundary conditions as Figure 9.6c, plus an impermeable top surface (outside the foundation). This case models a confined sand aquifer overlain by clay, in which a "water pocket" (see Section 9.12.2) may be created at the clay/sand plug interface. Since all water flow comes from the radial far field boundary (the flownet becomes rectangular), α_u increases by a factor 1.3 (now $\alpha_u \approx 0.32$, was $\alpha_u \approx 0.24$ for Figure 9.6c). Surprisingly, the decrease in dimensionless flow parameter F is only 15% (now $F \approx 0.42$, was $F \approx 0.50$). This is for steady state flow. The following commentary includes diffusion analyses of these last two cases and shows that their excess pore pressure–time curves differ little.

These findings are also applicable to other L/D values. Hence, using infinite half space solutions and uniform permeability for unconfined sand (i.e. the first three Figure 9.6 situations) are cautious – this case under-predicts α_u and, as shown in the example that follows, also under-predicts soil plug average seepage gradient i_{avg} and underpressure Δu. Since unit skin friction and unit end-bearing decrease with increasing i_{avg}, penetration resistance R is over-predicted.

As an example, consider a foundation with embedment ratio L/D = 1, L = 10 m and sand $\gamma_{sub} = 11$ kN/m³ (i.e. $i_{crit} = 1.1$). Using the infinite half space solution and uniform permeability $k_f = 1$ (i.e. the Figure 9.6a flownet), the $\alpha_u \approx 0.15$ value gives $i_{avg} = 0.85$ (8.5 m head loss/10 m length) for an underpressure $\Delta u = 100$ kPa. Since the hydraulic gradient i_{avg} (0.85) does not exceed i_{crit} (1.1), Δu can be increased. When i_{avg} equals i_{crit}, the critical underpressure Δu_{crit} value is ≈ 130 kPa (100 kPa × 1.1/0.85). If the sand plug permeability increases to, say, $k_f = 3$, then the results given on Figure 9.6b are applicable, i.e. $\alpha_u \approx 0.33$ (was 0.15). The Δu_{crit} value is now ≈ 165 kPa (100 kPa × 1.1/0.67) instead of 130 kPa. This represents a 25% Δu_{crit} increase – provided that a factor 3 sand plug permeability increase can be justified.

Regarding penetration resistance R, and defining penetration resistance multiplication factor $\eta_R = (1 - i_{avg}/i_{crit})$, consider the preceding example with underpressure Δu = 100 kPa. Assuming k_r = 1, we have i_{avg} = 0.85, giving η_R = (1 − 0.85/1.1) = 0.22. If in fact k_r = 3, then $\alpha_u \approx 0.33$ gives i_{avg} = 0.67 (6.7 m head loss/10 m length) for the same underpressure Δu = 100 kPa. This gives η_R = (1 − 0.67/1.1) = 0.39. Most sand seepage models assume a linear relationship between R and i_{avg}, i.e. R = 0 when η_R = 0. If so, then, for the previous example, R has been over-predicted by $\approx 75\%$ (η_R = 0.22 for k_r = 1, but actually η_R = 0.39 if k_r = 3).

Values of α_u and associated underpressure Δu are generally used to assess inner friction and tip resistance reduction, and, as shown earlier, i_{avg} needs to be less than i_{crit}, the critical underpressure. In addition, available underpressure Δu must exceed the underpressure required to overcome the penetration resistance R; see Section 9.11.2. Note that i_{crit} is usually not i_{avg}, but the exit seepage gradient (i.e. at seafloor), more details of which are given in Section 9.11.1. Penetration resistance R methods were summarised earlier in Section 9.8.1. For water flow, the relevant equation is $\Delta u = (R - W_{steel})/A_i$, where R is a function of Δu.

9.8.4.1 Commentary – steady state

Conventional design practice assumes steady state (not transient) conditions to analyse the pore pressure conditions in the soil plug and at caisson tip level. The following paragraphs assess how quickly steady state (equilibrium) is achieved and discuss the reasonableness of the steady state flow assumption.

Consider 100 kPa suction pressure (total stress decrease $\Delta\sigma$) instantaneously applied to the seafloor inside a caisson. At time t = 0, underpressures Δu created in the soil are equal to $\Delta\sigma$. However, $\Delta\sigma$ varies throughout the soil mass. A reasonable approximation is that Δu = 100 kPa throughout the soil sand plug within the caisson, diminishes with depth below caisson tip level using a 1:2 (h:v) load spread model (e.g. Δu = 16 kPa at 25 m bsf), and Δu = 0 elsewhere. With time, the Δu values dissipate to seafloor (Δu = 100 kPa and 0 kPa inside and outside respectively). Linear elastic soil and laminar flow are assumed. The transient flow 2D axisymmetric diffusion equation was solved using a slightly modified version of finite element program P80 (Smith and Griffiths, 1998). The following parameter values were adopted:

L caisson embedded length 10 m
D caisson diameter 10 m
c_v soil coefficient of consolidation 5 m²/s

L and D are both reasonable/typical caisson in sand values (L/D = 1)

Soil consolidation coefficient c_v was derived assuming:

k soil permeability 1e–3 m/s
q_c CPT resistance 10 MPa
D soil confined modulus $\approx$ 5 q_c (50e3 kPa)
γ_{water} water unit weight 10 kN/m³
c_v = k D/γ_{water} = 1e–3 × 50e3/10 5 m²/s ($\approx$ 1.5·10e8 m²/y).

Program P80 was used to re-analyse the four steady state cases shown on
Figure 9.6 (i.e. uniform, k_r = 3, impermeable and aquifer). Figure 9.7 presents normalised caisson tip Δu pore pressure ratio at tip level (i.e. α_u),
together with the corresponding t_{50} and t_{90} data points. The steady state (t =
200 s) α_u results are slightly higher than the Figure 9.6 α_u values. This is
probably due to program P72/Figure 9.6 averaging inner and outer nodal
Δu values, whereas program P80/Figure 9.7 used inner only.
    Figure 9.7 shows that:

1. elapsed times vary between 0.03 s and 0.06 s (t_{50}) and 3.5 s and 6.4 s
   (t_{90}) to achieve 50% and 90% of steady state Δu values
2. the fastest dissipation (lowest t_{90}) is 3.5 s for case (b) plug (k or) c_v:
   mass (k or) c_v ratio = 3
3. the slowest dissipation (highest t_{90}) is 6.4 s for case (a) uniform
   permeability
4. surprisingly, the Δu – t curves for cases (c) and (d) differ little.

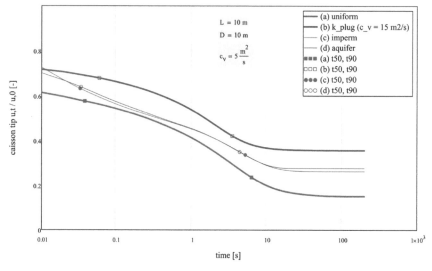

Figure 9.7 Caisson tip excess pore pressure – time. Axisymmetric unsteady state
fluid flow. Foundation L = D = 10 m. Soil mass c_v = 5 m²/s for cases (a)
through (d). Soil plug c_v = 15 m²/s for case (b).

In all four cases, steady state conditions are achieved rapidly (t_{90} within 6 seconds). This implies that (for the given geometry and parameter values) the steady state assumption is very reasonable. Using steady state Δu values is slightly cautious, since they are only marginally underpredicted. In addition, using case (a) is also cautious, since it has the "slowest" response.

To put this time (6 seconds) into perspective, assuming ≈ 1 mm/s penetration rate and 50 mm tip wall thickness WT, the caisson tip will have penetrated another 6 mm, i.e. just over 8% of WT.

Dissipation times (t_{90}) are essentially proportional to $1/c_v$ and L^2. The main source of error in c_v ($c_v = k D/\gamma_{water}$) is the uncertainty in soil permeability k (typically 2 or 3 orders of magnitude), followed by uncertainty in constrained modulus D (up to 1 order of magnitude). Note that, as the soil plug seepage gradient increases, k increases (soil becomes more permeable), and D decreases (soil becomes softer), which may possibly keep c_v more or less constant.

9.8.5 Classical bearing capacity method

Penetration resistance R is calculated using the equation:

$$R = Q_d = Q_f + Q_p = fA_s + qA_p \tag{9.8}$$

where

Q_f = skin friction resistance along sides of caisson and any protuberances

Q_p = total end-bearing resistance at caisson tip and any protuberances

f   = unit skin friction resistance

A_s = pile inside and outside surface areas embedded in soil

q   = unit end-bearing resistance

A_p = caisson tip cross-sectional area (excluding contained soil).

For undrained soil response, inner and outer unit skin friction, f, at any point along the foundation may be calculated by the equation

$$f = \alpha s_u \tag{9.9}$$

where

α = a dimensionless factor (see Tables 9.3, 9.4).

Unit end-bearing q may be computed by the equation

$$q = N_c s_u \tag{9.10}$$

where

N_c = bearing capacity factor using Table 9.5. This is from ISO 19901-4 (ISO 2016), not API 2GEO (API 2011), because end-bearing needs to be more carefully considered: the contribution is larger for penetrating caissons than for frictional plugged pipe piles.

Table 9.4 Recommended Adhesion α Factors.

Purpose	Soil/pile adhesion factor α [–]	Remarks
Installation, pull-out and retrieval (time t = 0)	α = 0.90 on $s_{u,r}$ (defined by Best Estimate shear strength s_u and sensitivities S_t)	
Pull-out (time t = ∞)	α = 0.70 (self-weight penetration zone)	(NGI, 1999)
	α = 0.65 (suction assistance zone)	
Removal (time t = ∞)	α = 0.90 on $s_{u,peak}$	

Notes: DNV (2005) recommends slightly different (usually higher) α values, e.g. API RP2GEO (2011)
Main Text for pull-out – but they are for driven (not pushed) piles
Pull-out excludes twist considerations, for which α should be decreased slightly

Table 9.5 Recommended N_c Factors (ISO 19901-4:2016).

Purpose	Shape of Area	Bearing Capacity Factor N_c [–]
Calculation of pile tip penetration resistance	Strip	7.5
Calculation of critical underpressure causing soil plug failure	Circular	6.2 to 9.0 depending on embedment ratio L/D $N_c \approx min[6.2 (1 + 0.2 \, L/D), 9]$
Calculation of penetration resistance of protuberances	Varies	5 to 13.5

For frictional soils, unit skin friction, f, may be calculated by the equation:

$$f = b\sigma'_{vo} \qquad (9.11)$$

where

β   = dimensionless skin friction factor

σ'_{vo} = effective overburden pressure at the depth in question.

Since coring intermediate foundations are assumed similar to driven open-ended pipe piles unplugged, Table 9.6 may be used for selection of β values if other data are not available. Unit end-bearing on the pile wall tip (and any protuberances) q_w may be computed using the equation:

$$q_w = N_q\sigma'_{vo,tip} \qquad (9.12)$$

where

$\sigma'_{vo,tip}$ = soil effective in-situ vertical stress at foundation tip level

N_q = dimensionless bearing capacity factor (see Table 9.6).

Table 9.6 Penetration Resistance in Silica Sand. Skin Friction and End-Bearing Parameters (ISO 19902:2007).

Relative density	D_r [%]	Soil type	Skin friction factor β [–]	Limit unit skin friction f_{lim} [kPa]	End-bearing factor N_q [–]	Limit unit end-bearing q_{lim} [MPa]
Very loose	0–15	Sand				
Loose	15–35	Sand				
Loose	15–35	Sand-silt	n/a	n/a	n/a	n/a
Medium dense	35–65	Silt				
dense	65–85	Silt				
Dense						
Medium dense	35–65	Sand-silt	0.29	67	12	3
Medium dense	35–65	Sand	0.37	81	20	5
Dense	65–85	Sand-silt				
Dense	65–85	Sand	0.46	96	40	10
Very dense	> 85	Sand-silt				
Very dense	> 85	Sand	0.56	115	50	12

Notes:
n/a : not applicable
Unit skin friction f = min(β σ'_{vo}, f_{lim})
Unit wall end-bearing q_w = min(N_q σ'_{vo}, q_{lim})

Intermediate suction foundations are usually not long enough to warrant limiting f and q_w to the limit values given in Table 9.6. In addition, the tabulated N_q values are for circles, not strips. Hence they over predict coring penetration resistance.

9.9 LANDING ON SEAFLOOR, MINIMUM SELF-WEIGHT PENETRATION AND FREE-FALL PENETRATION

9.9.1 Landing on seafloor

When landing on a sandy seabed, local scouring at the edge of the foundation and piping around the skirts are almost inevitable unless the lowering velocity is very low and the areas for water evacuation through the foundation base are large.

Section 7.3.7.6 of DNVGL-RP-C212 (DNV-GL, 2017, 2019) discusses piping when landing on sandy seabeds:

Open piping channels towards the end of the penetration shall be avoided in particular if final penetration by suction is necessary. The process of stopping the piping is complex and is not possible to analyse with a reasonable accuracy. It involves initially scouring and transport of particles horizontally on the seabed. As the penetration increases the transportation of the particles involves vertical lifting that requires higher velocities. Piping is likely to take place at local positions along the periphery corresponding to the lower spots of the seabed, whereas along the remaining parts of the periphery a penetration resistance will gradually build up and cause the piping to stop at some stage. There appears to be no available reliable method to analyse this process, so empirical data should be sought after and used for assessment of the potential for scouring and piping. Surveys after installation are important as a basis for deciding whether mitigations become necessary.

In addition, Section 6.2 of DNV-RP-H103 (DNV, 2014b) gives pertinent advice for analysis and modelling impact landing.

For controlled lowering, a reasonable foundation velocity just above seafloor (excluding vessel motions) is 0.2 m/s. This is because most design bases for shallow and intermediate foundations specify that the velocity should not exceed 0.2 m/s. This value also is implied in both ISO-19901-4 (ISO, 2016a) (Annex A.7.6.2.1) and API 2GEO (API, 2011) - Sections 7.14 and A.7.14 discusses installation effects. Both documents are for shallow (not intermediate) foundations. This (maximum) velocity of 0.2 m/s is a starting point for:

- assessing foundation drainage (shallow foundations)
- designing relief valves (suction caissons) – piping/soil plug base failure (push-out).

Velocities higher than 0.2 m/s occur during uncontrolled self-weight penetration (both caissons and open-ended pipe piles), and their free-fall penetration is discussed in Section 9.9.3.

Top plate vent design and suction pump design are discussed in Sections 12.6 and 12.7 respectively.

9.9.2 Minimum self-weight penetration

Self-weight penetration should generally be at least 0.5 m below seafloor all the way around the foundation tip perimeter. This is in order to:

- preclude local piping around the foundation tip
- start the suction assisted installation stage.

Occasionally, when competent soils (e.g. dense sand and very stiff clay) are present at seafloor, this penetration value may not be achieved.

For a given foundation geometry, possible measures to increase self-weight penetration are

- decrease foundation D and/or WT (i.e. reduce tip resistance)
- increase preload (e.g. tank or steel pipe pile filled with water, rock, concrete)

- make sure seafloor is as flat as possible (i.e. minimise risk of hydraulic leaks)
- add loose sand (before installation, perimeter footprint)
- bevel tip (i.e. attempt to reduce tip resistance)
- add sandbags around outer perimeter (after installation, dubious effectiveness)
- relocate intermediate foundation
- jetting (difficult to control).

Friction breakers and paint are unlikely to be effective in sand. This is because most of the self-weight penetration resistance is from end-bearing (not skin friction) resistance.

9.9.3 Free-fall penetration

Free-fall penetration occurs without crane assistance and a heavy foundation, and can also occur with a decreasing soil resistance ("hard over soft") profile. Since penetration is uncontrolled, risks need to be minimised. High free-fall penetration values are associated with heavy intermediate and pile foundations in clay. There is no free-fall risk during the subsequent suction assistance stage. This is because suction pumps operate at a semi-fixed flow rate; hence penetration is displacement, not load, controlled.

Uncontrolled penetration generally exceeds controlled self-weight penetration – this effect is used to install torpedo piles well below seafloor. Free-fall is best modelled using Newton's Second Law, forward Euler integration with time and Archimedes' Principle for buoyancy effects. Assuming "coring" in clay and ignoring viscous drag effects on the wall tip, the corresponding pseudo code to model free-fall in soil is:

```
program free_fall
initialise pile velocity and tip penetration depth;
subscripts 0 and 1 refer to previous and current timesteps
v₁ = max(v₀, 1e⁻³ m/s)
z₁ = 0 m/s
Δt = 0.01 s
loop while pile velocity v₁ > 0:
    v₀ = v₁
    z₁ = z₁ + v₀Δt
    SF: skin friction resistance by integrating sᵤ from 0 to
        z₁
    EB: end-bearing resistance using sᵤ at z₁
    SF₁ = ∫ Perim α sᵤ,z dz
    EB₁ = A_wall sᵤ,z1 N_c,strip,deep
    RF: vent (nozzle) resistance using v₀
```

```
      VN : vent (nozzle) velocity
      ΔH : vent differential head loss
      Δu : overpressure increase
      v_pile = v_0
      v_n = v_pile A_i/A_n
      ΔH = [1/(2 g)] [v_n²/C_d²]
      Δu = ΔH γ_water
  RF_1 = Δu (A_i - A_n)
  BF: buoyancy force at z_1, assuming constant wall area
  BF_1 = V_steel [(L - z_1) /L] γ_water + V_steel [z_1/L] γ_soil
  F: pile downwards force at z_1
  a, v: pile acceleration and velocity at z_1
  F_1 = M g - (SF_1 + EB_1 + RF_1 + BF_1)
  a_1 = F_1/M
  v_1 = v_0 + a_1Δt
end loop
exit with pile penetration z_1
```

The above pseudo code is easily modified to include free-fall in the water column above seafloor, and viscous effects. Viscosity is velocity dependent, and various torpedo pile researchers have proposed dimensionless drag coefficients, both above and below seafloor velocity dependent and shaft/tip viscosity models (e.g. O'Loughlin et al., 2004). For a given vent area, the vent resistance R_n is a function of the nozzle velocity and inverse discharge coefficient squared, i.e. $R_n \alpha (v_n/C_d)^2$. Section 12.6 (on top plate vent design) derives the R_n equation and also gives a worked example.

Both the free-fall model and 1-D drivability (wave equation using lumped masses and springs) analyse pile dynamics using integration with time. It is noted that pile buoyancy effects are missing from the original Smith (1960) drivability model, but it is possible in GRLWEAP (Pile Dynamics, 2010) to include pile buoyancy effects by changing the pile gravity. Office drivability practice generally excludes buoyancy – all piles are submerged, whether on land or offshore. Including buoyancy would make piles slightly harder to drive; this is because the additional upwards resistance increases soil resistance to driving (SRD).

9.9.3.1 Example – pile free-fall

To illustrate the difference a vent makes to penetration, two free-fall cases are analysed in a normally consolidated clay profile. Both have the same foundation geometry (D = 5 m, L/D = 4) and mass (67 tonne), and have the same seafloor velocity (v_0 = 0.2 m/s). Case A is a suction caisson with two number 1.0 m diameter vents, open during free-fall. Case B is an OWT monopile with an open top, i.e. without a top plate. Due to vent resistance, it is expected that Case A (suction pile) will penetrate less than Case B (OWT monopile).

The data, which are subsequently re-used in the Section 12.6 (Top Plate Vent design) example, for both cases are:

Pile: $D = 5$ m, $L = 20$ m, $WT_{side}/D = 200$, $WT_{top}/D = 100$, $\rho_{steel} = 7.7$ tonne/m³
$(D_i = 4.95$ m, Perim $= \pi\,(D + D_i) = 31.259$ m, $A_{wall} = 0.391$ m², $A_i = 19.244$ m²)
$(V_{steel} = 8.796$ m³, mass $M = 67.733$ tonne, weight $M\,g = 664.231$ kN)
Case A (suction foundation):   two vents, diameter $= 1.0$ m
$A_n = 0.8$ m²
Case B (OWT monopile):   no vent, vent diameter $= 4.95$ m
$A_n = 19.244$ m²

Soil: $s_u = 2 + 1.5\,z$ [kPa, m], $\rho_{soil} = 1.6$ tonne/m³, $\rho_{water} = 1.0$ tonne/m³
Pile – Soil – Fluid:

$$\alpha_i = \alpha_o = 0.3, N_{c,strip,deep} = 7.5$$
$$C_d = 0.61, v_0 = 0.2 \text{ m/s}.$$

Figure 9.8 compares the resulting Case A (coloured red/thick) and Case B (blue/thin) acceleration and velocity data versus depth. The following paragraphs discuss the free-fall and self-weight penetration results.

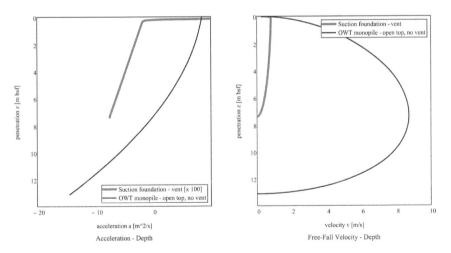

Figure 9.8 Free-fall penetration example: (a) foundation acceleration – depth and (b) velocity – depth. Foundation L = 20 m, D = 5 m, M = 67.7 tonne. Seafloor velocity = 0.2 m/s. Free-fall penetrations: 7.41 m (suction foundation with top plate vent) and 13.12 m (open top OWT monopile). Self-weight penetration = 7.37 m. Suction foundation accelerations have been increased by a factor 100.

9.9.3.2 Example – pile free-fall case A (suction foundation)

When released just above seafloor with velocity v_0 = 0.2 m/s, the suction foundation penetrates to 7.41 m bsf. The red/thick line on Figure 9.8a shows the free-fall acceleration – depth data (1524 time steps, 15.2 s). For clarity, accelerations have been multiplied by 100. It is seen that, just below seafloor, there is an initial acceleration of around 0.02 m/s². This is caused by the downwards acting foundation weight (664 kN) exceeding the upwards acting buoyancy force (86 kN) and vent resistance (143 kN at v_{pile} = 0.2 m/s), and essentially zero skin friction and end-bearing resistance close to seafloor. Below around 0.2 m depth, acceleration magnitudes remain small, reaching around –0.02 m/s² at final penetration. There is no acceleration, only a deceleration whose magnitude increases essentially linearly with depth. This linear decrease is due to both skin-friction and end-bearing resistance increasing almost linearly with depth, whereas buoyancy and vent resistances remain almost constant. The red/thick line on Figure 9.8b shows the corresponding free-fall velocity – depth data. The maximum velocity is just over 0.4 m/s (around twice v_0) near seafloor and the velocity decreases non-linearly until it eventually becomes zero at 7.41 m depth. This penetration depth is unlikely to be radically different had viscous effects been included – velocities are low. Similarly, penetration is also insensitive to seafloor velocity: the same value (7.41 m) is assessed for v_0 = 2.0 m/s. This is due to the vents slowing the foundation down.

9.9.3.3 Example – pile free-fall case B (OWT monopile)

For uncontrolled lowering (i.e. free-fall), released at seafloor (again with velocity v_0 = 0.2 m/s), the intermediate foundation will penetrate to 13.1 m bsf.

The blue/thin line on Figure 9.8a shows the acceleration – depth data (239 time steps, 2.4 s). It is seen that, unlike the suction foundation, acceleration is significantly higher – almost 1 g at seafloor. This is due to zero vent resistance. Acceleration reduces with penetration and becomes zero at around 7 m bsf, and thereafter begins to decelerate, reaching around –15 m/s² (just under 2 g) at final penetration.

As expected, due to the higher accelerations, the blue/thin line on Figure 9.8b shows that velocities are also higher, reaching a peak velocity of approximately 8.7 m/s. Theoretically, the 20 m long pile would be fully embedded if v_0 is around 20 m/s. However, because viscous effects have been excluded, this depth is unlikely to be achieved in the field.

9.9.3.4 Example – self-weight penetration Cases A and B

Since the foundation mass is constant, self-weight penetrations are identical for both cases. For controlled lowering (i.e. using a crane or similar), we have that skin friction (SF) + end-bearing resistance (EB) = pile buoyant weight. This can be written in pseudo-code notation as:

$$SF + EB + BF = M\,g.$$

The three terms on the left-hand side – SF, EB and BF (buoyancy force) – are all penetration depth dependent, and BF decreases slightly with increasing penetration into a higher-density material than water. By trial and error it is found that the SWP is 7.37 m. This value can be checked by verifying that the forces balance. At 7.37 m depth, we have SF = 520.6 kN, EB = 38.3 kN, BF = 105.3 kN, giving SF + EB + BF = 664.2 kN. As expected, this balances the 67.733 tonne pile, whose weight M g = 664.2 kN.

9.9.3.5 Example – commentary

For this foundation, the Case B (OWT monopile) free-fall value (13.1 m) is just over 175% of the self-weight penetration (7.51 m). Incorporating viscous drag effects are likely to reduce the Case B free-fall value. The Case A (suction foundation) free-fall value (7.41 m) is only 40 mm deeper than the self-weight penetration (7.37 m).

9.10 INSTALLATION IN CLAY

The key design challenge for installation analyses in clay is usually calculation of installation and allowable underpressure. The effect on allowable underpressure when encountering firm to stiff clay layers overlying sand and weaker clay layers can also be a challenge.

The allowable underpressure may be limited by maximum suction pump pressure (usually atmospheric or possibly vacuum, see Section 12.7) and structural considerations (steel cylinder buckling, etc.).

9.10.1 Base failure in clay

If the top plate vent is not properly designed, then excess water pressure build-up (overpressure) can occur during self-weight penetration. If significant, then conventional end-bearing failure at tip level is possible. Section 12.6 gives some top vent design considerations. Assessing maximum allowable overpressure is similar to that for underpressure, which is described in the remaining paragraphs.

Conversely, if the underpressure is too high, then geotechnical failure (reverse end-bearing) of the soil plug at foundation tip level can occur and should be checked. Various assessment methods are possible.

Maximum allowable underpressure (Δu_{all}) profiles, derived using static equilibrium of the clay soil plug, can be given by:

$$\Delta u_{all} = F_i / A_i + N_{c,circle} \; s_{u,av,tip} / \gamma_m \qquad (9.13)$$

where

F_i = inner skin friction resistance (= $\alpha \; s_{u,av} \; A_{int,surface}$)

A_i = plan view inside area where underpressure is applied

$N_{c,circle}$ = circular foundation bearing capacity factor $\approx$ min[6 (1 + 0.2 L/D), 9]

$s_{u,av,tip}$ = average $s_{u,z}$ between caisson tip (L) and depth (L + D $\alpha_{D,su}$)

$\alpha_{D,su}$ = s_u averaging parameter ($\alpha_{D,su} \approx 0.25$)

γ_m = material factor on end-bearing = 1.5

Suction anchor embedment ratios L/D are usually 6 or less for normally consolidated clays (e.g. Andersen et al., 2005). Equation (9.13), which is sensitive to inner friction adhesion parameter α_i, usually provides L/D values well in excess of 6.

ISO 19901-4 (ISO 2016, equation A.77) also uses static equilibrium but recommends a minimum safety factor of 1.5 on Δu_{all}, i.e. on both REB and F_i. Consequently, ISO 19901-4 (ISO 2016) gives lower Δu_{all} values (and lower limit embedment ratios) than DNV RP-E303 (DNV, 2018).

Note that using either the Skempton (1951) or the DNV (2005) equation $N_{c,circle}$ is likely to be cautious (i.e. they underestimate Δu_{all}) and provide an additional reserve on material factor γ_m. Section 10.5.2 gives reasons for $N_{c,circle}$ underestimation.

Randolph and Gourvenec (2011) considered the soil plug stability ratio (the ratio of underpressure to cause plug base failure to the underpressure required to penetrate the caisson to that depth). They present design curves of plug stability ratio and embedment ratio, along with an equation for limiting embedment ratio in normally consolidated soil conditions, as shown in Figure 9.9.

Section 9.10.3 gives an installation example for clay using the DNV RP-E303 (2005) equation. The pseudo code may be easily modified to include other base failure options.

At the other end of the scale, low embedment ratios (L/D $\approx$ 0.5) have not yet been investigated in detail. This is a typical potential problem in competent clays, where installation (rather than in-place capacity) governs. Conventional reverse end-bearing (REB) theory using $N_{c,circle}$ = 6 may not be sufficiently cautious. Numerical 3D FEA may be required to verify that the REB failure model is indeed applicable. The problem may be in fact non-axisymmetric. A possibility is an essentially circular/spheroidal failure arc, beginning inside the caisson and subjected to underpressure (i.e. negative

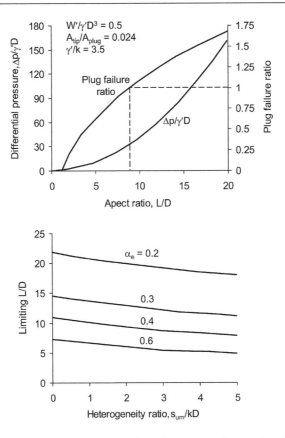

Figure 9.9 Design curves for suction foundation installation in clay (Randolph and Gourvenec, 2011).

surcharge), intersecting the caisson at two points on the caisson tip, and exiting at seafloor outside. Research may be required.

9.10.2 Plug heave in clay

Intermediate foundation penetration causes heave of the inner clay soil plug at seafloor to match part of the soil volume displaced by the foundation wall and the full volume of any internal stiffeners. Conventional practice is to assume a 50%–50% split during self-weight penetration but up to a 100%–0% split in favour of inward flow during suction assisted penetration (DNV, 2018).

Additional heave should be allowed for if significant free-standing soil is expected (e.g. above ring stiffeners), or underpressures are higher than Δu_{all} (maximum allowable underpressure).

The former is because a mixture of water and clay can be trapped between internal ring stiffeners, and that trapped water can give a significant contribution to soil plug heave (Andersen et al., 2005).

The latter is because the amount of soil entering the intermediate foundation is likely to be strongly affected by the proximity to internal base failure. During the early stages of suction assisted penetration, the factor of safety against base failure is high, and soil flow around the foundation tip is considered to be similar to that during the self-weight penetration stage. However, near the end of penetration, the soil plug is closer to failure requiring mobilisation of reverse end-bearing resistance for stability, and hence a greater proportion of soil met at the foundation tip is likely to flow inside.

Since installation will fail if the soil plug meets the top plate before the intermediate foundation has penetrated to target depth, the foundation overall height should be increased by at least the height of the plug heave. The additional height is equal to (part of) the foundation steel volume divided by its footprint area, plus the full volume of any internal stiffeners. For a 50%–50% split during both self-weight penetration and suction-assisted penetration stages, and no internal stiffeners, the additional height is given by $\Delta H_{heave}/L \approx 2\, WT/D$ (Romp, 2013). Hence, for wall thickness/diameter ratios between 1/300 and 1/100, height increases could be up to 2% of the embedded length. For large intermediate foundations (D = 5 m, L = 30 m), ΔH_{heave} could be around 0.75 m. More soil plug heave can be accommodated by domed (than flat) top plates.

9.10.3 Clay installation/retrieval example

To provide example resistance/underpressure–penetration depth diagrams, the classical bearing capacity method for clay ($f = \alpha s_u$ and $q = N_c s_u$) has been programmed to provide a simple clay installation (and retrieval/removal) model. It can be modified/improved to accommodate non-uniform foundation geometry, multiple clay layers and different opinions.

Assumptions include:

uniform caisson diameter D and wall thickness WT (i.e. no protuberances)

reverse wall tip end-bearing (i.e. same resistance during installation and retrieval)

DNV RP-E303 (2005) base failure equation (γ_m = 1.5) during suction-assisted installation.

Using these assumptions, the equations to find penetration resistances $R_{suction}$, $R_{retrieval}$ and corresponding underpressure $\Delta u_{suction}$, retrieval overpressure $\Delta u_{retrieval}$ and allowable underpressure Δu_{all} at any tip penetration L are:

$$R_{suction} = F_i + F_o + Q_w \tag{9.14}$$

$$R_{\text{retrieval}} = R_{\text{suction}} \tag{9.15}$$

$$\Delta u_{\text{suction}} = \left(R_{\text{suction}} - W_{\text{sub,steel}} \right) / A_{\text{plug}} \tag{9.16}$$

$$\Delta u_{\text{retrieval}} = \left(R_{\text{retrieval}} + W_{\text{sub,steel}} \right) / A_{\text{plug}} \tag{9.17}$$

$$\Delta u_{\text{all,suction}} = F_i / A_{\text{plug}} + N_{\text{c,circle}} \, s_{u,\text{av,tip}} / \gamma_{m,\text{install}} \tag{9.13, 9.18}$$

$$\Delta u_{\text{all,retrieval}} = F_i / A_{\text{plug}} + N_{\text{c,circle}} \, s_{u,\text{av,tip}} / \gamma_{m,\text{retrieval}} \tag{9.19}$$

where

$$F_i = A_{si}\alpha_i \int s_{u,z} dz \tag{9.20}$$

$$F_o = A_{so}\alpha_o \int s_{u,z} q_{c,z} dz \tag{9.21}$$

$$Q_w = A_{\text{wall}} N_{\text{c,strip,deep}} \, s_{u,\text{av,tip}} \tag{9.22}$$

and

α_i	= clay-steel inner adhesion factor (unit friction $f_i = \alpha_i s_u$)
α_o	= clay-steel outer adhesion factor (unit friction $f_o = \alpha_o s_u$)
A_{plug}	= caisson internal (suction) and clay plug area
	= $\pi D_i^2/4$
A_{si}	= caisson inner perimeter
A_{so}	= caisson outer perimeter
A_{wall}	= caisson wall tip area
L	= caisson tip penetration into clay
L_{can}	= caisson total length
$N_{\text{c,strip}}$	= strip foundation bearing capacity factor (clay)
	$\approx \min[5\,(1 + 0.2\,L/D),\ 7.5]$
$N_{\text{c,strip,deep}}$	= deep strip foundation bearing capacity factor (clay) ≈ 7.5
R	= soil resistance
$s_{u,\text{av,tip}}$	= average $s_{u,z}$ between caisson tip (L) and depth ($L + D_i\,\alpha_{D,su}$) (note: D_i, not D)
$\alpha_{D,su}$	= s_u averaging parameter ($\alpha_{D,su} \approx 0.25$)
$W_{\text{sub,steel}}$	= caisson submerged weight (including ballast)

The corresponding pseudo code for the above installation in clay is:

```
program install_clay
initialise and evaluate constants:
D_i = - 2 WT, A_si = π D_i, A_so = π D, A_p = π (D² - D_i²) /4 and
     A_plug = π D_i²/4
```

```
ΔL = 0.1 m
nL = Lcan/ΔL + 1
zero matrix ANS (size nL rows by 12 columns)
k = 0
```

loop tip penetration L from 0 to Lcan in ΔL steps:

```
k = k + 1
L = (k - 1) ΔL
Fi: inner skin friction resistance by integrating su
    from 0 to L
Fo: outer skin friction resistance by integrating su
    from 0 to L
```

$F_i = A_{si} \alpha_i \int s_{u,z} dz$ and $F_{o,clay} = A_{so} \alpha_o \int s_{u,z} dz$

Q_w: end-bearing resistance using s_u at depth L

$Q_w = A_{wall} N_{c,strip,deep} s_{u,L}$

$R_{suction} = F_i + F_o + Q_w$

$R_{retrieval} = R_{suction}$

$\Delta u_{suction} = (R_{suction} - W_{sub,steel})/A_{plug}$

$\Delta u_{retrieval} = (R_{retrieval} + W_{sub,steel})/A_{plug}$

$s_{u,av,tip}$ = average $s_{u,z}$ between caisson tip (L) and depth (L + $D_i \alpha_{D,su}$)

$\Delta u_{all,install} = F_i/A_{plug} + N_{c,circle} s_{u,av,tip}/\gamma_{m,install}$

$\Delta u_{all,rerievall} = F_i/A_{plug} + N_{c,circle} s_{u,av,tip}/\gamma_{m,retrieval}$

store in k^{th} row of matrix ANS:

k, L, and L/L_{can} in columns 1 through 3

F_i, F_o, Q_w, $R_{suction}$ and $R_{retrieval}$ in columns 4 through 8

$\Delta u_{all,install}$ and $\Delta u_{suction}$ in columns 9 and 10

$\Delta u_{all,retrieval}$ and $\Delta u_{retrieval}$ in columns 11 and 12

```
end loop on tip penetration L
exit with matrix ANS
```

matrix ANS (nL rows, 12 columns) contents:

col# 01: k	matrix ANS row number	[–]
col# 02: L	caisson tip penetration into sand	[m]
col# 03: L / D	caisson penetration/diameter ratio	[–]
col# 04: F_i	caisson inner wall friction resistance	[kN]
col# 05: F_o	caisson outer wall friction resistance	[kN]
col# 06: Q_w	caisson wall tip resistance	[kN]
col# 07: $R_{suction}$	soil resistance during penetration	[MN]
col# 08: $R_{retrieval}$	soil resistance during retrieval	[MN]
col# 09: $\Delta u_{all,install}$	allowable water (under) pressure	[kPa]
col# 10: $\Delta u_{suction}$	water (under) pressure during penetration	[kPa]
col# 11: $\Delta u_{all,retrieval}$	allowable water (over) pressure	[kPa]
col# 12: $\Delta u_{retrieval}$	water (over) pressure during retrieval	[kPa]

Installation plots:

col# 07 ($R_{suction}$) versus col# 02 (L) *installation resistance – depth*

col# 10 ($\Delta u_{suction}$) and col# 09 ($\Delta u_{all,install}$) versus col# 02 (L) *underpressure – depth*

Retrieval plots:

col# 08 ($R_{retrieval}$) versus col# 02 (L) *retrieval resistance – depth*

col# 12 ($\Delta u_{retrieval}$) and col# 11 ($\Delta u_{all,retrievall}$) versus col# 02 (L) *overpressure – depth*

9.10.3.1 Example – clay installation and retrieval

Consider a typical lightly overconsolidated clay ($s_u = 5 + 2\,z$) in which a 4 m diameter D anchor pile caisson has to be installed to 17 m below seafloor (bsf) using self-weight and suction assistance. Check for base failure during suction assisted phase using $\gamma_m = 1.5$ on plug base heave failure. Also assess (immediate, time t = 0) retrieval, but with a slightly lower γ_m value (say 1.25) to assess overpressure

```
Pile: D = 4 m, L_can = 17 m, WT = D/200 = 20 mm, W_sub,steel
      = 750 kN
      (D_i = 3.96 m, A_s,i = π D_i = 12.44 m, A_s,o = π D =
      12.57 m, A_wall = 0.25 m², A_plug = 12.32 m²)
Soil: clay, s_u,z = 5 + 2 z [kPa, m]
Pile - Soil:
α_i = α_o = 0.30
α_D,su = 0.25
γ_m,install = 1.5 and γ_m,retrieval = 1.25.
```

Using the data just presented, at the end of penetration (L = 17 m), we have

k   = 171
L   = 17 m
L/D = 4.25

and

$s_{u,av,L}$ = average s_u from 0 to L (17 m) = s_u at L/2 (8.5 m) = 5 kPa + 2 kPa/m × 8.5 m = 22 kPa

$\int s_{u,z}\,dz$ = average s_u from 0 to L (17 m) × L = 22 kPa × 17 m = 374 kN/m

$s_{u,tip}$ = s_u at L (17 m) = 5 kPa + 2 kPa/m × 17 m = 39 kPa

$s_{u,av,tip}$ = average s_u from L to L + 0.25 D_i (17 m to 17.99 m) = 40 kPa

and

F_i $= A_{si}\, \alpha_i \int s_{u,z}\, dz = 12.44 \text{ m} \times 0.30 \times 374 \text{ kN/m} = 1396 \text{ kN}$

F_o $= A_{so}\, \alpha_o \int s_{u,z}\, dz = 12.57 \text{ m} \times 0.30 \times 374 \text{ kN/m} = 1410 \text{ kN}$

Q_w $= A_{wall}\, N_{c,strip,deep}\, s_{u,tip} = 0.25 \text{ m}^2 \times 7.5 \times 39 \text{ kPa} = 73.1 \text{ kN}$

$R_{suction}$ $= F_i + F_o + Q_w = 1396 \text{ kN} + 1410 \text{ kN} + 73.1 \text{ kN} = 2.88 \text{ MN}$

$R_{retrieval}$ $= R_{suction} = 2.88 \text{ MN}$

$\Delta u_{suction}$ $= (R_{suction} - W_{sub,steel})\, /A_{plug} = (2.88 \text{ MN} - 0.75 \text{ MN})\, /12.32 \text{ m}^2 =$
173 kPa

$\Delta u_{retrieval}$ $= (R_{retrieval} + W_{sub,steel})\, /A_{plug} = (2.88 \text{ MN} + 0.75 \text{ MN})\, /12.32 \text{ m}^2 =$
295 kPa

$\Delta u_{all,install}$ $= F_i/A_{plug} + N_{c,circle}\, s_{u,av,tip}/\gamma_{m,install}$
$= 1396 \text{ kN}/12.32 \text{ m}^2 + 9.0 \times 40 \text{ kPa} /1.5 = 113 \text{ kPa} + 240 \text{ kPa} =$
353 kPa

$\Delta u_{all,retrieval}$ $= F_i/A_{plug} + N_{c,circle}\, s_{u,av,tip}/\gamma_{m,retrieval}$
$= 1396 \text{ kN}/12.32 \text{ m}^2 + 9.0 \times 40 \text{ kPa} /1.25 = 113 \text{ kPa} + 288 \text{ kPa}$
$= 401 \text{ kPa}$

Figure 9.10 shows two resistance and pressure profiles. Figure 9.10a is for penetration. It is seen that, at final tip penetration (17 m bsf), the required underpressure $\Delta u_{suction}$ (blue/thin line) is $\approx$ 175 kPa. The corresponding allowable suction pressure Δu_{all} (red/thick line) is just over 350 kPa. Since Δu_{all} exceeds $\Delta u_{suction}$, there is no major risk of soil plug base failure during suction assisted penetration. The self-weight penetration value is 7.6 m, and is more than sufficient to start the suction assisted stage. The high self-weight penetration (nearly half the embedded pile length) is not uncommon for anchor piles in weak clay, and self-weight penetration would decrease for α_i and α_o values in excess of 0.3.

Figure 9.10b shows the corresponding retrieval profiles. These were obtained using the same $\alpha_i = \alpha_o = 0.3$ value and the same soil s_u profile, but with a $\gamma_m = 1.25$ (was 1.5). Because there is (reverse) end-bearing, the retrieval and penetration resistance profiles are identical. Unlike penetration, retrieval resistance (and overpressure) drops once the pile starts to move upwards. Hence the overpressure profiles (both required and allowable) are valid only at/near the penetration depth at which retrieval commences. Since pile submerged weight opposes retrieval (instead of assisting penetration), the overpressure magnitude is higher than the corresponding underpressure: at 17 m bsf, $\Delta u_{retrieval}$ is now 295 kPa (compared to $\Delta u_{suction}$ = 173 kPa). Despite the higher pressure, there is still no base failure risk at all depths, except apparently above 3 m bsf. This is unlikely to occur – this zone is well above the SWP depth (7.6 m bsf), and a vessel crane is used to keep the suction anchor vertical and lift the it from the seafloor.

9.10.3.2 Commentary

Interested readers are encouraged to perform sensitivity analyses. For a given s_u profile, key input variables include α_i, α_o and $W_{sub,steel}$. The model assumes a uniform pile geometry. In practice, anchor piles are rarely

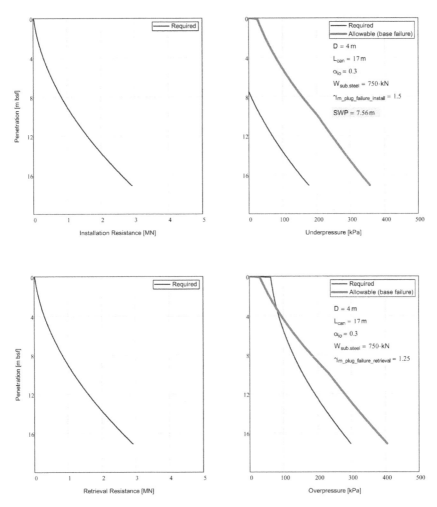

Figure 9.10 Installation and retrieval resistance in clay. Example using f = αs$_u$ method. Anchor pile D = 4 m, L = 17 m in LOC clay (s$_u$ = 5 + 2z [kPa, m]).

uniform: internal ring stiffeners (usually an increased WT at/near lug level) are used to transfer chain loads, and inner protuberances are used to reduce the cylinder WT and steel stresses. In such cases, α$_i$ (and WT) is no longer constant and has to be depth dependent. The pseudocode is capable of handling such modifications.

At the time of writing there is no (ISO, DNV-GL) advice for selecting a suitable γ$_m$ value for overpressure assisted retrieval. The example presented here used a lower γ$_m$ value (1.25) for retrieval than installation, i.e. 1.5 according to DNV RP-E303 (2005). This lower γ$_{m,retrieval}$ value implies that the probability of plug base failure is lower during retrieval because of different design parameters. The main reason is usually adopting higher (HE or

BE) α values for retrieval and lower (LE or BE) α values for installation. Another reason may be that, during the preliminary design phase, when both installation and retrieval assessments are made, installation data are available from previous projects nearby, thereby reducing the uncertainty bandwidth. Other organisations may have other $\gamma_{m,retrieval}$ opinions. In any case, $\gamma_{m,retrieval}$ values of less than 1.5 should be justified.

Since the tip resistance contribution is small, and unit friction $f = \alpha s_u$, it is obvious that various α and s_u combinations can be used to compute the same f (and hence assess resistances and pressures). In the example, s_u has been kept constant.

Even though there is limited risk of plug base failure during suction-assisted penetration, base failure could be possible during the preceding self-weight penetration stage if high overpressures occur in the "water plug" due to undersized vents. Section 12.6 (Top Plate Vent Design) below includes an example calculation.

9.10.4 Friction set-up in clay

Set-up is the increase of skin friction resistance with time in normally consolidated and lightly overconsolidated clays. Friction increases with time after installation due to thixotropy, dissipation of excess pore water pressures and soil consolidation. Due to their smaller displacement ratio and installation by suction, pore pressure dissipation is expected to occur more quickly for suction foundations than for driven pipe piles.

Friction resistance component is usually large compared to total end-bearing resistance. An "alpha" method (i.e. maximum friction is a proportion of the undrained shear strength) is employed, with different alpha values for the self-weight and suction assisted penetration zones. The shear strength reduction factor (α) at time $t = 0$ is generally close to the reciprocal of the soil sensitivity (S_t) value.

More information can be found in the following references:

ISO 19901-4:2016 Section A.8.1.3.2.5: changes in axial capacity in clay with time (piles)
ISO 19901-4:2016 Section A.11.5.2.2.4: increase of side friction with time (suction anchors)
DNV RP-E303 (2005) Sections 4.7 and 4.8: give outside and inside set-up factors for suction anchors. Since factor values are low estimates, they should be increased for retrieval and removal assessments.

For suction foundations in clay, two resistance estimates are usually provided at various time periods after installation:

- a Low Estimate for pull-out axial resistance
- a High Estimate for retrieval/removal.

Such friction set-up is for normally consolidated clays. Set-up magnitude will be less (and could possibly decrease) for over consolidated clays. This is due to the different pore water pressure regimes during and after installation.

9.10.5 Boulders in clay

Foundation tip integrity (buckling initiation) may be adversely affected by large-sized material. Such materials include boulders in glacial till, drop-stones in weak clays, coral rubble in carbonate clay, and flintstones in chalk. Like sand and gravel, boulder angularity can vary from angular (sharp edges with relatively plane sides with unpolished surfaces) to rounded (smoothly curved sides and no edges); see Figure 9.11.

This section outlines a reasonable model and design procedure should there be a possibility of encountering boulders (or similar) during suction foundation installation in clay. Assumptions made include (a) the boulder is spherical with diameter $D_{boulder}$ and (b) a constant coefficient of friction μ models the boulder–steel interface and (c) diameter $D_{boulder} < D$ should the boulder be within the pile. The design procedure consists of three stages: (i) check for refusal, then, if there is no refusal, (ii) check foundation tip integrity and modify foundation tip geometry if necessary. Then (iii), repeat stages (i) and (ii) if the geometry has changed. Since the forces acting on the pile tip are needed for all stages, these are first derived in the following paragraphs.

Figure 9.11 Boulders, Whiterose Development, Grand Banks, Offshore Newfoundland. PanGeo (2010). This area is unsuitable for suction foundations - ground conditions include hard clays and very dense granular glacial till with boulders.

9.10.5.1 Forces on boulder and pile tip

Figure 9.12a shows top and plan view of the boulder and the pile wall contact point. This point, which is usually offset from the boulder centreline, is defined by an angle θ_{xz} measured anti-clockwise from the horizontal. Hence the pile wall is precisely on the boulder centreline when $\theta_{xz} = 90°$, and the pile is just grazing the edge of the boulder when $\theta_{xz} = 0°$ (boulder inside the pile) or $180°$ (boulder outside the pile).

Figure 9.12b shows the corresponding forces acting on the boulder, both inside and outside the pile. R_s, the soil resistance may be calculated by standard bearing capacity formulae ($R_s = AN_c s_u$). Unlike shallow foundations, the boulder is "deep" (depth $> 4\ D_{boulder}$, say) and R_s is omnidirectional (i.e. independent of load inclination).

Knowing α_{xz}, R_s and $W_{sub,boulder}$, the unknown boulder force F_n and angle β_{xz} can be derived using the polygon of forces, and hence the required pile tip lateral and vertical forces F_x and F_z, and their resultant R_p, can be found (see Figure 9.13).

Pile tip forces F_x, F_z and R_p for any pile tip penetration L and angle θ_{xz} are given by the equations:

If $0 \le \theta_{xz} \le 90°$

$$F_x = F_n fac1 \tag{9.23}$$

$$F_z = F_n fac2 \tag{9.24}$$

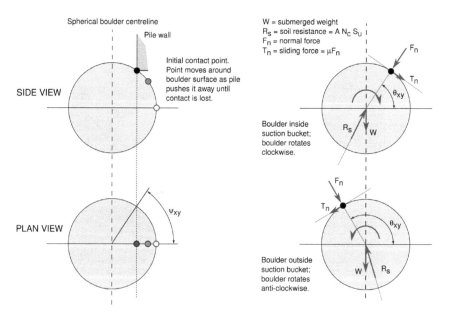

Figure 9.12 Boulder in clay – left, boulder and pile wall contact point, and right, forces acting on the boulder.

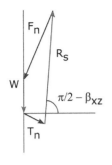

Boulder force polygon
Forces W, R_S and angle θ_{xz} known,
find unknown force F_n and angle β_{xz}

W = submerged weight
R_S = soil resistance = A N_C S_u
F_n = normal force
T_n = sliding force = μF_n

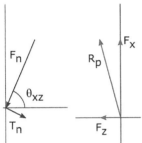

Pile tip forces
$F_x = F_n \cos(\theta_{xz}) - \mu \sin(\theta_{xz})$
$F_z = F_n \sin(\theta_{xz}) + \mu \cos(\theta_{xz})$

$R_p = (F_x^2 + F_z^2)^{0.5}$

Maximum $R_p = R_S - W$

Figure 9.13 Boulder in clay – boulder polygon of forces and equivalent pile tip forces.

If $90\,^\circ \leq \theta_{xz} \leq 180°$

$$F_x = F_n fac3 \tag{9.25}$$

$$F_z = F_n fac4 \tag{9.26}$$

where

$$fac1 = \cos(\theta_{xz}) - \mu \sin(\theta_{xz}) \tag{9.27}$$

$$fac2 = \sin(\theta_{xz}) + \mu \cos(\theta_{xz}) \tag{9.28}$$

$$fac3 = \cos(\gamma_{xz}) - \mu \sin(\gamma_{xz}) \tag{9.29}$$

$$fac4 = \sin(\gamma_{xz}) + \mu \cos(\gamma_{xz}) \tag{9.30}$$

$$\gamma_{xz} = 180° - \theta_{xz} \tag{9.31}$$

and

$$R_p = \sqrt{\left(F_x^{\;2} + F_z^{\;2} \right)} \tag{9.32}$$

and

$$F_n = \left\{ \sqrt{\left[R_s^{\;2} + \left(1+\mu^2\right) - \left(W_{sub,boulder}\; fac1\right)^2 \right]} - W_{sub,boulder}\; fac2 \right\} / \left(1+\mu^2\right) \tag{9.33}$$

$$T_n = \mu\, F_n \tag{9.34}$$

$$R_s = A_{boulder}\, N_{c,circle}\, s_{u,tip} \tag{9.35}$$

$$N_{c,circle} = \min\left[6\left(1+0.2\, L/D_{boulder}\right), 9 \right] \tag{9.36}$$

$$W_{sub,boulder} = V_{boulder}\, \gamma_{sub,boulder} \tag{9.37}$$

$$A_{boulder} = \pi\, D_{boulder}^{\;2}/4 \tag{9.38}$$

$$V_{boulder} = \left(4/3\right)\pi\left(D_{boulder}/2\right)^3 \tag{9.39}$$

Figure 9.14 shows how pile tip forces F_x, F_z and R_p vary with θ_{xz} (Example 1 following, L = 17 m, μ = 0.6), and Table 9.7 lists force variations for

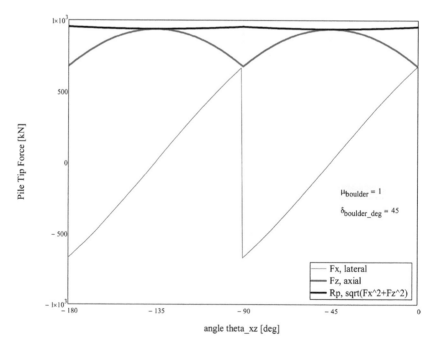

Figure 9.14 Boulder in clay – pile tip forces versus θ_{xz}. Example 1: D boulder = 2 m at L = 17 m.

Table 9.7 Boulder and Pile Tip Forces – Sensitivity to Friction Coefficient μ.

μ [-]	δ [deg]	$F_{x,max}$ [kN]	$F_{z,max}$ [kN]	$R_{p,min}$ [kN]	$R_{p,max}$ [kN]	$R_{p,max}/\Delta R_{boulder}$ [-]
0.2	11	963	937	937	982	1.048
0.4	22	903	938	937	972	1.037
0.6	31	827	938	938	964	1.028
0.8	39	749	938	937	958	1.022
1.0	45	674	937	937	953	1.017

Notation:
μ = boulder-pile friction coefficient
δ = boulder-pile friction angle, atan(μ)
$F_{x,max}$ = maximum pile tip lateral force
$F_{z,max}$ = maximum pile tip axial force
$R_{p,min}$ = minimum pile tip resultant force, $R_p = \sqrt{(F_x^2 + F_z^2)}$
$R_{p,max}$ = minimum pile tip resultant force, $R_p = \sqrt{(F_x^2 + F_z^2)}$
R_s = soil resistance on boulder, $N_c A_{boulder} s_u$
$W_{sub,boulder}$ = boulder submerged weight, $V_{boulder} \gamma_{sub,boulder}$
$\Delta R_{boulder}$ = net soil resistance on boulder, $R_s - W_{sub,boulder}$

Notes:
[1] Refusal Example 1 – weak clay with 2 m diameter dropstone
[2] R_s = 995kN
[3] $W_{sub,boulder}$ = 57 kN
[4] $\Delta R_{boulder} = R_s - W_{sub,boulder}$ = 938 kN

friction coefficient μ varying between 0.2 and 1.0. This figure and table are key items. Note that:

(a) F_x, F_z and R_p kinks at θ_{xz} = 90° are due to the sign change in T_n
(b) maximum lateral F_x has almost the same magnitude as maximum axial F_z. (Example 1: $F_{x,max}$ = 827 kN, $F_{z,max}$ = 938 kN)
(c) axial $F_z = F_n$ when θ_{xz} = 90°
(d) lateral $F_x = T_n$ when θ_{xz} = 0° and 180°
(e) maximum axial F_z occurs when θ_{xz} = atan(1/μ) (Example 1: 59° for μ = 0.6)
(f) resultant R_p magnitude is essentially constant (Example 1: $R_{p,min}$ = 938 kN, $R_{p,max}$ = 964 kN)
(g) maximum resultant $R_{p,max} \approx \Delta R_{boulder}$ (Example 1: $R_{p,max}$ = 964 kN, R_s = 995 kN, $W_{sub,boulder}$ = 57 kN, $\Delta R_{boulder} = R_s - W_{sub,boulder}$ = 938 kN, $R_{p,max}/\Delta R_{boulder}$ = 1.028)

Findings (f) and (g) make possible a reasonably accurate simplification for both refusal and integrity assessments. That is, $\Delta R_{boulder}$ (= $R_s - W_{sub,boulder}$) the net soil *resistance*, can be applied as a *force* $R_{p,max}$ on the pile tip at *any* angle $\theta_{xz'}$ – anywhere from vertical to horizontal. Since the force initiating lateral buckling is 50% of that for axial buckling (Aldridge et al., 2005), this means that lateral is more important than axial.

9.10.5.2 Refusal

Encountering boulders causes "spikes" in soil resistance–depth profiles. This is similar to CPTu encountering gravel in sand, where the q_c value temporarily increases as the obstruction is pushed aside. However, underpressure Δu increases whereas CPT u_2 decreases. Refusal occurs if Δu exceeds allowable $\Delta u_{all,install}$.

9.10.5.3 Refusal example 1 – weak clay with dropstone

Example 1 re-uses the installation data example given in Section 9.10.4. Check for pile refusal during suction-assisted phase should a 2 m diameter boulder (dropstone) be encountered within the LOC clay at final tip penetration. To maximise the axial resistance increase, assume that the initial point of contact is on top of the spherical boulder. Take *force* R_p = *resistance* $\Delta R_{boulder}$

Relevant data are:

Pile: $D = 4$ m, $L_{can} = 17$ m, WT = D/200 = 20 mm, $D_i = 3.96$ m, $A_{plug} = 12.32$ m², $W_{sub,steel} = 750$ kN
Soil: clay, $s_{u,z} = 5 + 2\,z$ [kPa, m]
Pile-Soil: $\alpha_i = \alpha_o = 0.30$
Boulder: $D_{boulder} = 2$ m ($A_{boulder} = 3.14$ m², $V_{boulder} = 4.19$ m³), $\gamma_{sub,boulder} = 16$ kN/m³ ($W_{sub,boulder} = 536$ kN), $\mu = 0.6$.

At the end of penetration (L = 17 m) and no boulder (from the Section 9.10.4 example), we have:

$R_{suction}$ = $F_i + F_o + Q_w$ = 1396 kN + 1410 kN + 73.1 kN = 2.88 MN
$\Delta u_{suction}$ = $(R_{suction} - W_{sub,steel})$ /A_{plug} = (2.88 MN – 0.75 MN) /12.32 m² = 173 kPa
$\Delta u_{all,install}$ = $F_i/A_{plug} + N_{c,circle}\,s_{u,av,tip}\,/\gamma_{m,install}$
= 1396 kN/12.32 m² + 9.0 × 40 kPa/1.5 = 113 kPa + 240 kPa = 353 kPa

For a 2 m diameter boulder at the pile tip, we have:

$s_{u,tip}$ = s_u at L (17 m) = 5 kPa + 2 kPa/m × 17 m = 39 kPa
$R_{boulder}$ = $N_{c,circle}\,s_{u,tip}\,A_{boulder}$ = 9.0 × 39 kPa × 3.14 m² = 1102 kN
$\Delta R_{boulder}$ = $R_{boulder} - W_{sub,boulder}$ = 1102 kN – 536 kN = 566 kN
$R_{suction,boulder}$ = $R_{suction} + \Delta R_{boulder}$ = 2.88 MN + 566 kN = 3.45 MN
$\Delta u_{suction,boulder}$ = $(R_{suction,boulder} - W_{sub,steel})$ /A_{plug} = (3.45 MN – 0.75 MN) /12.32 m² = 219 kPa

It is seen that, if a single boulder is met at final tip penetration (17 m bsf), the required underpressure $\Delta u_{suction,boulder}$ rises by 46 kPa to 219 kPa (was 173 kPa). The corresponding allowable suction pressure Δu_{all} was 353 kPa. Hence, the design is satisfactory – there is no refusal ($\Delta u_{suction,boulder} \leq \Delta u_{all}$). Had the boulder been encountered at a shallower depth, or the initial contact point was not directly on top of the boulder, then both $R_{boulder}$, and $\Delta u_{suction,boulder}$ would be smaller. Note that $\Delta R_{boulder}$, the net soil resistance, subtracts $W_{sub,boulder}$ from $R_{boulder}$.

Foundation tip integrity assessments will subsequently demonstrate that the tip WT has to be increased from 20 mm to at least 35 mm to preclude local buckling.

9.10.5.4 Refusal example 2 – competent clay/glacial till with boulders

Example 2 is similar to Example 1 but with three clay layers, 1 m of 10 kPa soft drape, underlain by 4 m of firm to stiff clay (s_u = 40/100 kPa), and very stiff 175 kPa glacial till containing occasional boulders below 5 m depth. The stiffer clays require a stubbier pile (D = 5 m, L/D = 2). Check for refusal should *two* 2 m diameter boulders be met at final tip penetration.

Pile: D = 5 m, L_{can} = 10 m, WT = D/100 = 50 mm, D_i = 4.9 m,
            A_{plug} = 18.86 m²,
            $W_{sub,steel}$ = 1750 kN
Soil: clay, $s_{u,z}$ = 10 kPa, z ≤ 1 m; $s_{u,z}$ = 40/100 kPa, 1 < z ≤ 5 m;
            $s_{u,z}$ = 175 kPa, z > 5 m
Pile-Soil: $\alpha_i = \alpha_o$ = 0.50
Boulder: same as Example 1.

Because both pile and soil data are different, a separate calculation using the Section 9.10.4 program is needed. The fine details are not presented. At target penetration depth (L = 10 m) and no boulder, we have:

$R_{suction}$ = $F_i + F_o + Q_w$ = 8.97 MN + 9.15 MN + 1.02 MN = 19.14 MN
$\Delta u_{suction}$ = $(R_{suction} - W_{sub,steel}) / A_{plug}$ = (19.14 MN – 1.75 MN) /18.86 m²
        = 0.92 MPa
$\Delta u_{all,install}$ = $F_i/A_{plug} + N_{c,circle} s_{u,av,tip} /\gamma_{m,install}$
        = 8.97 MN/18.86 m² + 8.5 × 175 kPa/1.5 = 0.47 MPa + 0.99 MPa
        = 1.46 MPa

For two boulders at the pile tip, we have:

$s_{u,tip}$ = $s_{u,z}$ at L (10 m) = 175 kPa
$R_{boulder}$ = $N_{c,circle} s_{u,tip} A_{boulder}$ = 8.5 × 175 kPa × 3.14 m² = 4.67 MN
$\Delta R_{boulder}$ = $R_{boulder} - W_{sub,boulder}$ = 4.67 MN – 536 kN = 4.14 MN

$R_{\text{suction,boulder}}$ $= R_{\text{suction}} + 2 \times \Delta R_{\text{boulder}} = 19.14 \text{ MN} + 2 \times 4.14 \text{ MN} = 27.42$ MN

$\Delta u_{\text{suction,boulder}} = (R_{\text{suction,boulder}} - W_{\text{sub,steel}}) / A_{\text{plug}} = (27.42 \text{ MN} - 1.75 \text{ MN}) / 18.86 \text{ m}^2 = 1.36 \text{ MPa}$

It is seen that, if two boulders are met at final tip penetration (10 m bsf), the required underpressure $\Delta u_{\text{suction,boulder}}$ is 1.36 MPa (was 0.92 MPa). The corresponding Δu_{all} value is 1.46 MPa. There is again no refusal ($\Delta u_{\text{suction,boulder}} \leq \Delta u_{\text{all}}$), but is marginal ($\Delta u_{\text{suction,boulder}}/\Delta u_{\text{all}} = 0.93$). This implies that $\gamma_{\text{m,install}}$ is closer than usual to, but still exceeds 1.5. Major challenges are likely had both boulders been larger than 2 m diameter. Foundation tip integrity assessments will subsequently demonstrate that, for the current pile and soil data, WT has to be increased from 50 mm to 100 mm to preclude local buckling and buckle propagation. Underpressures are extremely high, and structural steel checks are required. A domed top is necessary.

9.10.5.5 Foundation tip Integrity

If pile tip refusal is unlikely, then foundation steel tip integrity assessments should be made. According to the analytical methods of Aldridge et al. (2005), a local buckle will occur at the pile tip when the lateral force F_x or a near-axial force F_z values are:

$$F_x = 1.4\,\sigma_y WT^2 \tag{9.40}$$

$$F_z = 2.8\,\sigma_y WT^2 \tag{9.41}$$

where

$$\sigma_y = \text{pile yield stress}\left(\approx 345 \text{ MPa for steel}\right) \tag{9.42}$$

This local buckle will propagate if:

$$\left(D / WT\right)^3 < 5\left(1 - v_{\text{soil}}^2\right)E_{\text{pile}} / E_{\text{soil}} \tag{9.43}$$

where

$$v_{\text{soil}} = \text{soil Poisson's ratio}\left(= 0.5 \text{ for clay}\right) \tag{9.44}$$

$$E_{\text{pile}} = \text{pile Young's modulus}\left(\approx 210\text{e}3 \text{ MPa for steel}\right) \tag{9.45}$$

$$E_{\text{soil}} = \text{soil Young's modulus}\left(\approx 400\,s_u \text{ for clay}\right) \tag{9.46}$$

Since both examples do not refuse, pile tip integrity assessments should be made. Again, take forces F_x and F_z equal to net boulder resistance $\Delta R_{\text{boulder}}$. Example 1 and 2 $\Delta R_{\text{boulder}}$ values were 0.566 MN and 6.14 MN respectively.

Table 9.8 Pile Tip Integrity Assessment Examples.

Parameter	Units	Refusal Example 1 – weak clay with dropstone		Refusal Example 2 – competent clay/glacial till	
		Original WT	Revised WT	Original WT	Revised WT
D	[m]	4	4	5	5
WT	[mm]	20	35	50	100
WT/D	[–]	1/200	1/100	1/100	1/50
$s_{u,tip}$	[kPa]	39	39	175	175
$F_{x,boulder} = \Delta R_{boulder}$	[MN]	0.57	0.57	4.67 [1]	4.67
$F_{z,boulder} = \Delta R_{boulder}$	[MN]	0.57	0.57	4.67	4.67
Local Buckle – Equation (9.40) $F_{x,buckle} = 1.4\ \sigma_y\ WT^2$					
$F_{x,buckle}$	[MN]	0.19	0.59	1.21	4.83
Local x-buckle?	[Yes/No]	Yes	No	Yes	No
Local Buckle – Equation (9.41) $F_{z,buckle} = 2.8\ \sigma_y\ WT^2$					
$F_{z,buckle}$	[MN]	0.39	1.54	2.42	9.66
Local z-buckle?	[Yes/No]	Yes	No	Yes	No
Buckle Propagation – Equation (9.43) $(D/WT)^3 < 5\ (1 - \nu_{soil}^2)\ E_{pile}/E_{soil}$					
$(D/WT)^3$	[–]	8000e3	n/a	1000e3	n/a
$5(1 - \nu_{soil}^2)E_{pile}/E_{soil}$	[–]	50.5e3	n/a	7.9e3	n/a
Buckle propagation?	[Yes/No]	Yes	n/a	Yes	n/a

Notation:
n/a not applicable - no local buckling
Notes:
[1] point load from a single boulder
[2] pile: σ_y = 345 MPa, E_{pile} = 210e3 MPa
[3] soil: E_{soil} = 400 $s_{u,tip}$, ν_{soil} = 0.5

If local buckling occurs, then increment WT in 5 mm steps until buckling stops to find the minimum required WT. Table 9.8 presents results. It is seen that:

(a) Example 1. Buckling (and propagation) occurs with 20 mm WT: $F_{x,boulder}$ (0.57 MN) is larger than $F_{x,buckle}$ (0.19 MN). However, with a revised WT value of 35 mm, there is no local buckling: $F_{x,boulder}$ (0.57 MN) no longer exceeds $F_{x,buckle}$ (0.59 MN). Propagation is not applicable – there is no buckling.

(b) Example 2 has the same outcome as Example 1: the original WT (50 mm) needs to be increased to at least 100 mm, where $F_{x,boulder}$ (4.67 MN) is less than $F_{x,buckle}$ (4.83 MN).

9.10.5.6 Commentary – suction pile

Boulders in sand cause suction pile refusal. Other installation methods (e.g. impact driving) are necessary. Hence there is no comparable text to be found in Section 9.11 (Installation in Sand).

The friction coefficient μ value (whether static or sliding), and where the point of contact is on the boulder, are both irrelevant: it is reasonable to assume that the R_p magnitude remains essentially the same. It is also considered sufficiently accurate to use the net force ($\Delta R_{boulder} = R_s - W_{sub,boulder}$) for both axial and lateral integrity assessments.

The model described here has assumed that the bearing capacity factor of a sphere equals $N_{c,circle}$. Ball penetrometer research indicates higher N_c values for deep (full flow) conditions, namely $11 < N_{c,ball} < 15$ depending on the interface α (0 – 1) value (Randolph et al., 2000). No solutions are available for "wished-in-place" near-surface spheres at shallow depth, but N_c values are also expected to be higher than those for shallow circular foundations. As usual, opinions differ, but $6 < N_{c,circle} < 9$ used herein is considered reasonable: major model inaccuracies include assumptions of shape (e.g. an ellipse instead of sphere) and ULS (the boulder is pushed aside before achieving the peak soil resistance).

In practice, since the geometry has changed for both the examples, Stage (iii) (= Stage (i) re-check for refusal followed by Stage (ii) reassess pile tip integrity) should be carried out. Stage (i) implies additional "no boulder" installation analyses with (a) increased $W_{sub,steel}$ and (b) increased WT values. However, increasing $W_{sub,steel}$ further increases Δu_{all}, which is beneficial. The "no boulder" soil resistance $R_{suction}$ value will not increase markedly, since the Q_{wall} increase will be offset by a smaller F_i value. It is also advisable to perform analyses with other alpha values (α_{LE}, α_{BE}, α_{HE}), and to check that there is no risk of over-penetration (particularly for the weak clay Example 1) due to increased foundation weight, $W_{sub,steel}$.

9.10.5.7 Commentary – anchor chain

Since the anchor chain will contact a (spherical) boulder if the boulder is within a distance $D_{boulder}/2$ of the chain edge, V_{chain}, the soil volume swept out by a (straight, constant θ_{lug}) chain, is given by:

$$V_{chain} = A_{chain} W_{chain} \qquad (9.47)$$

where

$$A_{chain} = cross-sectional \; area$$
$$= (1/2)\left[z_{lug} / \tan(volume) \right]\left[z_{lug} + D_{boulder} / 2 \right] \qquad (9.48)$$

$$W_{chain} = plan \; width = D_{chain} + D_{boulder} \qquad (9.49)$$

$$D_{chain} = chain \; effective \; width \; in \; bearing$$
$$= 2.5 \times nominal \; chain \; diameter_r \qquad (9.50)$$

V_{chain} may be compared with soil volumes swept out by an anchor pile. Plugged implies that the soil plug contains internal stiffeners/protuberances, whereas coring is for an essentially constant WT pile with no pad-eye stiffeners etc. Values of $V_{pile,plugged}$ and $V_{pile,coring}$ are given by:

$$V_{pile,plugged} = (\pi / 4)(D + D_{boulder})^2 (L + D_{boulder} / 2) \qquad (9.51)$$

$$V_{pile,coring} = (\pi / 4)\left[(D + D_{boulder})^2 - (D - D_{boulder})^2 \right](L + D_{boulder} / 2) \quad (9.52)$$

Typically, $V_{chain} \approx 0.5\, V_{pile,coring}$, but, as usual, swept volumes are sensitive to pile, chain and boulder geometry. Nevertheless, assuming boulders are randomly distributed in 3D space, then, even though the probability of boulder contact is greater for the pile than the chain, due diligence implies that the chain–boulder contact should also be analysed. The major issue is that θ_{lug} will increase, decreasing the FOS on axial pull-out failure. Whatever analysis option is selected, the crux is to select parameter values giving the required p (soil normal/bearing resistance per unit chain length) and f (soil frictional resistance per unit chain length) (Vivatrat et al., 1982) and to uniformly distribute p and f over an embedded chain length, ΔL_{chain} modelling the contact zone. A reasonable ΔL_{chain} value is $D_{boulder} / 4$. Using the pile tip model, $p = (R_s - W_{sub,boulder}) / \Delta L_{chain}$ and $f = \mu p$. Analysis options include (a) introducing a competent soil layer at the boulder depth and (b) modifying the chain analysis program. These options are left to interested engineers to investigate.

9.11 INSTALLATION IN SAND

9.11.1 Sand plug liquefaction, piping and heave in sand

Liquefaction, piping and heave are closely interlinked. Underpressure creates an axial upwards flow/seepage in the sand plug. When this is sufficiently high, inner friction and tip resistance have both been sufficiently reduced and penetration occurs. If the underpressure exceeds that causing liquefaction and/or piping, then penetration is halted and installation fails (liquefaction possibly results in sand in the suction pump and piping drops the underpressure to zero, and resistance increases). Due to the decreased effective stress within the sand plug, soil expansion occurs, and this causes additional sand plug heave. Additional side-effects of seepage/heave include increased permeability and reduced strength of the soil plug, both due to an increased voids ratio. All are discussed in more detail here.

Hence, the key design challenges for installing intermediate foundations in sand are:

(a) to sufficiently reduce inner friction and foundation tip resistance such that penetration can occur without incurring

(b) liquefaction failure of sand plug
(c) piping failure around foundation wall
(d) excessive sand plug heave

and, if in shallow water,

(e) the maximum underpressure should not exceed the lowest astronomical tide (LAT) water depth (Section 9.7).

Ad (a) reduced inner friction and foundation tip resistance. Unlike clay,
most resistance is from the foundation tip, and the primary effect of underpressure is to reduce this tip resistance rather than to increase the load
(which is a secondary effect).

Ad (b) sand plug liquefaction failure. Underpressure is traditionally limited to that which gives zero vertical effective stress, i.e. seafloor i_{crit} (critical
hydraulic gradient) $= \gamma_{sub}/\gamma_{water}$.

The foregoing implies that underpressure cannot exceed this critical value.
However, there is some evidence from carefully executed laboratory permeability tests (e.g. Fleshman and Rice, 2014; Panagoulias, 2016; Panagoulias
et al., 2017), and from foundation installation, both in the field and centrifuge (summarised by Panagoulias, 2016; Panagoulias et al., 2017) that i >
i_{crit} in medium-dense to very dense sands. Figure 9.15 shows suction

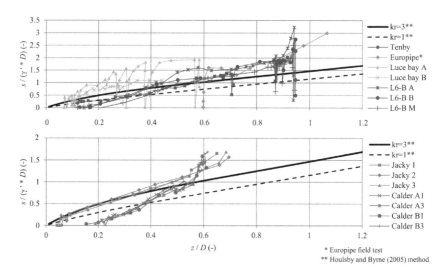

Figure 9.15 Normalised suction penetration curves (Panagoulias et al., 2017)
(With permission of SUT, London)

penetration data for six North Sea locations, together with the theoretical (Houlsby and Byrne, 2005) i_{crit} curves. No sand plug liquefaction occurred. In addition, no excessive sand plug heave was reported in any of the six cases considered. Hence, provided that due care is taken, gross failure is unlikely at underpressures (slightly) exceeding this critical hydraulic gradient. However, at the time of writing, a credible model has yet to be published, and installation procedures have not yet been generally accepted.

Upward seepage, even if at pressures less than critical, also creates some loosening/density reduction within the soil plug (Tran, 2005). This leads to internal sand heave, which, if not allowed for by stick-up/over height, may prevent the foundation penetrating to its design depth. In addition, Tran's small-scale model tests showed that no major adverse hydraulic conditions (only 10% heave) occurred, even with L/D ratios up to 2.0 and critical gradient > 1. Another factor mitigating severe sand plug liquefaction is probably passive arching in aged sands – arching theory (for the no-flow case) has been given by Paikowsky (1989).

The loosening/density reduction effect needs to be accounted for in in-place resistance assessments – the sand plug strength is reduced from original in-situ values.

Ad (c) piping failure. More important than heave is piping. If sand "boiling" occurs, piping channels may be locally formed, especially along the wall/soil plug interface. These will create a hydraulic leak – the water plug beneath the top plate is in direct contact with the seawater via a continuous channel along the foundation inner and outer wall. In this case, the suction pump is merely removing seawater. There is no water flow within the soil to reduce inner friction and end-bearing resistance. Also, a mixture of sand and water may come into the suction pump. In this case, installation will fail.

From the very limited amount of data available, it appears that piping failure is more likely when the foundation is restrained from penetration. The basic reason is that the foundation tip can no longer penetrate into weakened soil. Hence, obstructions in sand, such as cobbles or boulders, can quickly initiate local piping.

Ad (d) excessive soil plug heave. This is greater in sand than in clay (for a given foundation geometry). This is because there is an additional component of sand expansion due to stress decreases from upwards flow and other effects. If high installation pressures are anticipated, and domed top plates are required, these can take more soil plug heave than flat top plates. In the Gorm Field caisson installation (Senpere and Auvergne, 1982) excessive sand heave and piping failure was observed. Excess sand had to be removed by water jetting to enable the caissons to reach design penetration depth. No installation details were released – e.g. underpressure magnitudes.

Ad (e) maximum underpressure should not exceed the LAT water depth. This is a potential problem only in shallow water. European examples include the southern North Sea, Irish Sea and the Baltic.

Due diligence/mitigation measures for minimising the risk of liquefaction, piping, soil plug heave, etc., include:

- instrumentation
- careful installation
- retrieval/reinstallation
- pump flowrate (as high as possible, e.g. Tran, 2005)
- increased preload (gives smaller underpressure)
- reverse flow (i.e. overpressure) at end of installation. This is similar to hydraulic sand fill placement causing compaction. Reverse flow pushes back the sand plug and increases plug relative density/strength, but a potential disadvantage is that it increases/creates a void under baseplate.

9.11.2 Models for sand

As noted earlier, the effect of water flow has to be considered, and penetration resistance methods can be either the CPT q_c or beta type (see Section 9.8.1). A variety of analytical and empirical models have been developed and applied to assess installation resistance in sand. At the time of writing, there is no single industry-wide accepted method. More information can be found in the following references (in chronological order):

Erbrich and Tjelta (1999)	Statoil/Kvaerner	beta
Houlsby and Byrne (2005)	Oxford	beta
Andersen et al. (2008)	NGI	beta and CPT q_c
Senders and Randolph (2009)	UWA	CPT q_c
Chatzivasileiou (2014)	SPT	CPT q_c
Sturm et al. (2015)	NGI	beta.

Installation data for fewer than 20 projects are in the public domain (e.g. Panagoulias, 2016; Alderlieste and Van Blaaderen, 2015). Hence, the models described here (and associated parameter values) are likely to evolve/ improve as more data become available and additional back-analysis is performed. In addition, attempts are made to numerically investigate this phenomenon; see e.g. Martinelli et al. (2020).

Figure 9.16 compares critical underpressure as a function of embedment ratio based on published equations. Only two (Houlsby and Byrne, 2005; Sturm et al., 2015) account for a lower soil plug permeability. All equations are for a water flow source at seafloor (not for a source at a near-infinite radius), i.e. they cannot be used for confined sand aquifers.

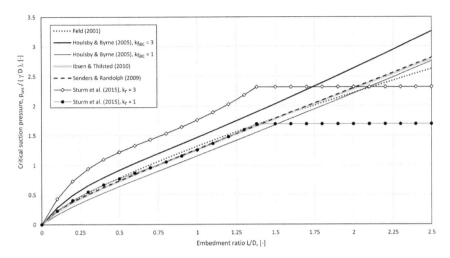

Figure 9.16 Comparison critical suction pressures (based on Senders and Randolph, 2009). Critical underpressure, $\Delta u_{crit}/(\gamma'D)$ versus embedment ratio L/D, k_r: inner soil plug/outer soil permeability ratio, [–].

9.11.3 Sand installation example

To provide example installation resistance/underpressure–penetration depth diagrams, the DNV CPT q_c resistance equations can be programmed to provide a simple model for sand installation. The resulting model is capable of modifications/improvements to accommodate different opinions.

Assumptions are numerous and include:

uniform caisson D and WT (i.e. no protuberances)

no silo effect within sand plug

DNV CPT q_c method (i.e. $q_w = k_p q_c$ and $f = k_f q_c$)

uniform permeability sand (i.e. no sand plug loosening)

outer skin friction: no increase due to flow

inner skin friction and wall tip end-bearing: linear decrease proportional to $1 - i_{sand}$

critical Δu based on average (not exit) hydraulic gradient i_{sand}

Houlsby and Byrne (2005) α_u – L/D equation.

Using these assumptions, the equations to find penetration resistance $R_{suction}$ and corresponding underpressure $\Delta u_{suction}$ at any tip penetration L are:

$$R_{suction} = F_{i,suction} + F_{o,suction} + Q_{w,suction} \qquad (9.53)$$

$$\Delta u_{suction} = \left(R_{suction} - W_{sub,steel}\right) / A_{plug} \qquad (9.54)$$

where

$$F_{i,\ suction} = F_{i,no\_flow}\left(1 - i_{sand}\right) \qquad (9.55)$$

$$F_{o,suction} = F_{o,no\_flow} \qquad (9.56)$$

$$Q_{w,suction} = Q_{w,no\_flow}\left(1 - i_{sand}\right) \qquad (9.57)$$

$$F_{i,no\_flow} = A_{si}k_{f,sand}\int q_{c,z}dz \qquad (9.58)$$

$$F_{o,no\_flow} = A_{so}k_{f,sand}\int q_{c,z}dz \qquad (9.59)$$

$$Q_{w,no\_flow} = A_{wall}k_{p,sand}q_{c,L} \qquad (9.60)$$

$$i_{sand} = \left(1 - \alpha_u\right)\Delta u_{suction} / \left(\gamma_{sub}L\right) \qquad (9.61)$$

$$\alpha_u = 0.45 - 0.36\left[1 - \exp\{\left(L/D\right)/0.48\}\right] \qquad (9.62)$$

Since iteration is required to balance loads and resistances, we have two more equations:

$$L_{suction} = A_{plug}\Delta u_{suction} + W_{sub,steel} \qquad (9.63)$$

$$R_{suction} = L_{suction} \qquad (9.64)$$

where, noting that subscripts "suction", "liq" and "no_flow" mean "with suction assistance and fluid flow", "required for full sand plug liquefaction" and "with suction assistance but no fluid flow" respectively:

α_u = excess pore pressure ratio at tip level

          = $\Delta u_{suction}$ (tip)/$\Delta u_{suction}$

Δu_{no_flow} = water (under) pressure – no fluid flow

Δu_{liq} = water (under) pressure – full liquefaction

$\Delta u_{suction}$ = water (under) pressure – fluid flow

γ_{sub} = sand plug submerged unit weight

γ_{water} = (sea) water unit weight

A_{plug} = soil plug area

A_p = caisson wall tip area

A_{si} = caisson inner perimeter

A_{so} = caisson outer perimeter

D = caisson outer diameter

D_i = caisson inner diameter

F_{i,no_flow} = caisson inner wall friction resistance – no fluid flow

$F_{i,liq}$ = caisson inner wall friction resistance – full liquefaction

$F_{i,suction}$ = caisson inner wall friction resistance – fluid flow

F_{o,no_flow} = caisson outer wall friction resistance – no fluid flow

$F_{o,liq}$ = caisson outer wall friction resistance – full liquefaction

$F_{o,suction}$ = caisson outer wall friction resistance – fluid flow

i_{sand} = sand plug (vertical) average seepage gradient

$k_{f,sand}$ = DNV empirical coefficient – unit skin friction resistance in sand

$k_{p,sand}$ = DNV empirical coefficient – unit end-bearing resistance in sand

L = caisson tip penetration into sand

L_{can} = caisson total length

$L_{suction}$ = load on caisson steel – fluid flow

$q_{c,z}$ = CPT cone tip resistance at depth z below seafloor

$q_{c,L}$ = CPT cone tip resistance at caisson tip depth L

Q_{w,no_flow} = caisson wall tip resistance – no fluid flow

$Q_{w,liq}$ = caisson wall tip resistance – full liquefaction

$Q_{w,suction}$ = caisson wall tip resistance – fluid flow

R_{no_flow} = soil resistance – no fluid flow

R_{liq} = soil resistance – full liquefaction

$R_{suction}$ = soil resistance – fluid flow

$W_{sub,steel}$ = caisson submerged weight (including ballast)

WT = caisson wall thickness

The corresponding pseudo code for the installation in the sand model is:

```
function αu_fun: evaluate αᵤ at any caisson tip penetration
    L and diameter D
ref Houlsby and Byrne, 2005
in: L, D // out: αᵤ
    c0 = 0.45, c1 = 0.36 and c2 = 0.48
    αᵤ = c0 - c1 [1 - exp {(-L/D) /c2}]
end
```

function *isand_fun*: evaluate i_{sand} at any caisson tip penetration L, diameter D, underpressure Δu and soil submerged unit weight γ_{sub}

```
in: L, D, Δu, γ_sub // out: i_sand
    αᵤ = αᵤ_fun(L, D)
    i_sand = [1 - αᵤ Δu/ (γ_sub L) ]
    i_sand = min(i_sand, 1)
end
```

```
program install_sand
initialise and evaluate constants
Dᵢ = D - 2 WT, Aₛᵢ = π Dᵢ, Aₛₒ = π D, Aₚ = π (D² - Dᵢ²) /4 and
    Aₚₗᵤg = π Dᵢ² /4
ΔL = 0.1 m
nL = Lcan /ΔL + 1
zero matrix ANS (size nL rows by 21 columns)
k = 0
```

loop tip penetration L from 0 to Lcan in ΔL steps:

```
    k = k + 1
    L = (k - 1) ΔL
```
F_{i,no_flow}: inner skin friction resistance by integrating q_c
   from 0 to L
F_{o,no_flow}: outer skin friction resistance by integrating q_c
   from 0 to L
$F_{i,no\_flow} = A_{si} k_{f,sand} \int q_{c,z} \, dz$ and $F_{o,clay} = A_{so} k_{f,sand} \int q_{c,z} \, dz$
Q_{w,no_flow}: end-bearing resistance using q_c at depth L
$Q_{w,no\_flow} = A_{wall} k_{p,sand} q_{c,L}$
$R_{no\_flow} = F_{i,no\_flow} + F_{o,no\_flow} + Q_{w,no\_flow}$
$\Delta u_{no\_flow} = (R_{no\_flow} - W_{sub,steel}) /A_{plug}$
$F_{i,liq} = 0$
$F_{o,liq} = F_{o,no\_flow}$
$Q_{w,liq} = 0$
$R_{liq} = F_{i,liq} + F_{o,liq} + Q_{w,liq}$
$\Delta u_{liq} = (R_{liq} - W_{sub,steel}) /A_{plug}$
$\alpha_u = au\_fun (L, D)$

iterate using $\Delta u_{suction}$ to balance resistance $R_{suction}$ and
   load $L_{suction}$
   $i_{sand} = isand\_fun (L, D, \Delta u_{suction}, \gamma_{sub})$
   $F_{i,suction} = F_{i,no\_flow} (1 - i_{sand})$
   $F_{o,suction} = F_{o,no\_flow}$
   $Q_{w,suction} = Q_{w,no\_flow} (1 - i_{sand})$
   $R_{suction} = F_{i,suction} + F_{o,suction} + Q_{w,suction}$
   $L_{suction} = A_{plug} \Delta u_{suction} + W_{sub\_steel}$
exit iteration with $\Delta u_{suction}$, etc., when $R_{suction} = L_{suction}$
store in kᵗʰ row of matrix ANS:
   k, L, and L/Lcan in columns 1 through 3
   F_{i,no_flow}, F_{o,no_flow}, Q_{w,no_flow}, R_{no_flow} and Δu_{no_flow} in columns 4
      through 8
   α_u and i_{sand} in columns 9 and 10
   $F_{i,liq}$, $F_{o,liq}$, $Q_{w,liq}$, R_{liq} and Δu_{liq} in columns 11 through 15
   $F_{i,suction}$, $F_{o,suction}$, $Q_{w,suction}$, $R_{suction}$, $L_{suction}$ and $\Delta u_{suction}$ in
      columns 16 through 21
```
```

```
end loop on tip penetration
exit with matrix ANS
```

matrix ANS (nL rows, 21 columns) contents:

col# 01: k	matrix ANS row number	[–]
col# 02: L	caisson tip penetration into sand	[m]
col# 03: L / D	caisson penetration/diameter ratio	[–]
col# 04: F_{i,no_flow}	caisson inner wall friction – no fluid flow	[kN]
col# 05: F_{o,no_flow}	caisson outer wall friction resistance – no fluid flow	[kN]
col# 06: Q_{w,no_flow}	caisson wall tip resistance – no fluid flow	[kN]
col# 07: R_{no_flow}	soil resistance – no fluid flow	[MN]
col# 08: Δu_{no_flow}	water (under) pressure – no fluid flow	[kPa]
col# 09: α_u	excess pore pressure ratio at tip level	[kN]
col# 10: i_{sand}	sand plug (vertical) average seepage gradient	[–]
col# 11: $F_{i,liq}$	caisson inner wall friction resistance – full liquefaction	[kN]
col# 12: $F_{o,liq}$	caisson outer wall friction resistance – full liquefaction	[kN]
col# 13: $Q_{w,liq}$	caisson wall tip resistance – full liquefaction	[kN]
col# 14: R_{liq}	soil resistance – full liquefaction	[MN]
col# 15: Δu_{liq}	water (under) pressure – full liquefaction	[kPa]
col# 16: $F_{i,suction}$	actual caisson inner wall friction resistance – fluid flow	[kN]
col# 17: $F_{o,suction}$	actual caisson outer wall friction resistance – fluid flow	[kN]
col# 18: $Q_{w,suction}$	actual caisson wall tip resistance – fluid flow	[kN]
col# 19: $R_{suction}$	actual soil resistance – fluid flow	[MN]
col# 20: $L_{suction}$	load on caisson steel – fluid flow	[MN]
col# 21: $\Delta u_{suction}$	actual water (under) pressure – fluid flow	[kPa]

Plot col# 07 (R_{no_flow}), col# 14 (R_{liq}), and col# 19 ($R_{suction}$) versus col# 02 (L) *resistance – depth*

Plot col# 08 (Δu_{no_flow}), col# 15 (Δu_{liq}), and col# 21 ($\Delta u_{suction}$) versus col# 02 (L) *underpressure – depth*

9.11.3.1 Example – sand installation

Consider a typical North Sea type profile (1.5 m of loose sand, $D_r \approx 0.3$, underlain by dense sand, $D_r \approx 0.8$) in which a 10 m diameter, 12 m length caisson has to be installed. Input data are:

Pile: D = 10 m, L_{can} = 12 m, WT = D/200 = 50 mm, $W_{sub,steel}$ = 6000 kN (D_i = 9.90 m, $A_{s,i} = \pi D_i$ = 31.10 m, $A_{s,o} = \pi D$ = 31.41 m, A_p = 1.565 m², A_{plug} = 77.0 m²)

Soil: sand 0 – 1.5 m: q_c = 2 z; >1.5 m: q_c = 10 + 1.5 z [MPa, m], γ_{sub} = 12 kN/m³

Water: γ_{water} = 10 kN/m³

Pile – Soil:

$$k_{p,sand}, k_{f,sand} = 0.3, 0.001 \text{ (BE) and } k_{p,sand}, k_{f,sand} = 0.6, 0.003 \text{ (HE)}$$

For the analysis using DNV q_c HE coefficients, midway through penetration, at L = 6 m, we have:

k = 61
L = 6 m
L/D = 0.6

and

$F_{i,no\_flow} = A_{si}\, k_{f,sand} \int q_{c,z}\, dz = 31.10 \text{ m} \times 0.003 \times 72.5 \text{ MN/m} = 6770 \text{ kN}$
$F_{o,no\_flow} = A_{so}\, k_{f,sand} \int q_{c,z}\, dz = 31.41 \text{ m} \times 0.003 \times 72.5 \text{ MN/m} = 6840 \text{ kN}$
$Q_{w,no\_flow} = A_{wall}\, k_{p,sand}\, q_{c,L} = 1.565 \text{ m}^2 \times 0.6 \times 19 \text{ MPa} = 17800 \text{ kN}$
$R_{no\_flow} = F_{i,no\_flow} + F_{o,no\_flow} + Q_{w,no\_flow} = 6770 \text{ kN} + 6840 \text{ kN} + 17800 \text{ kN}$
$\quad = 31.4 \text{ MN}$
$\Delta u_{no\_flow} = (R_{no\_flow} - W_{sub,steel})/A_{plug} = (31.4 \text{ MN} - 6 \text{ MN})/77.0 \text{ m}^2 = 330 \text{ kPa}$
$\alpha_u = c0 - c1\, [1 - \exp\{(-L/D)/c2\}] = 0.45 - 0.36\, [1 - \exp\{(-L/D)/0.48\}]$
$\quad = 0.193$
$\Delta u_{suction} = 72.1 \text{ kPa}$ *by iteration, verified below* $(R_{suction} = L_{suction})$
$i_{sand} = (1 - \alpha_u)\, \Delta u_{suction}/(\gamma_{sub}\, L) = 0.807 \times 72.1 \text{ kPa}/(12 \text{ kN/m}^3\, 6 \text{ m}) = 0.81$

and

$F_{i,liq} = 0 \text{ kN}$
$F_{o,liq} = F_{o,no\_flow} = 6840 \text{ kN}$
$Q_{w,liq} = 0 \text{ kN}$
$R_{liq} = F_{i,liq} + F_{o,liq} + Q_{w,liq} = 0 \text{ kN} + 6840 \text{ kN} + 0 \text{ kN} = 6.84 \text{ MN}$
$\Delta u_{liq} = (R_{liq} - W_{sub,steel})/A_{plug} = (6.84 \text{ MN} - 6 \text{ MN})/77.0 \text{ m}^2 = 10.9 \text{ kPa}$

and

$F_{i,suction} = F_{i,no\_flow}\, (1 - i_{sand}) = 6770 \text{ kN}\, (1 - 0.81) = 1300 \text{ kN}$
$F_{o,suction} = F_{o,no\_flow} = 6840 \text{ kN}$
$Q_{w,suction} = Q_{w,no\_flow}\, (1 - i_{sand}) = 17800 \text{ kN}\, (1 - 0.81) = 3420 \text{ kN}$
$R_{suction} = F_{i,suction} + F_{o,suction} + Q_{w,suction} = 1300 \text{ kN} + 6840 \text{ kN} + 3420 \text{ kN} =$
$\quad 11.5 \text{ MN}$
$L_{suction} = A_{plug}\, \Delta u_{suction} + W_{sub,steel} = 77.0 \text{ m}^2 \times 72.1 \text{ kPa} + 6000 \text{ kN} = 11.5 \text{ MN}$
$\Delta u_{suction} = (R_{suction} - W_{sub,steel})/A_{plug} = (11.5 \text{ MN} - 6 \text{ MN})/77.0 \text{ m}^2 = 72.1 \text{ kPa}$

Figure 9.17 shows results for 10 m diameter caissons tipping out in dense sand ($D_r \approx 80\%$). Figure 9.17a is for Best Estimate DNV k_p and k_f values. It is seen that, if there is no flow, the Best Estimate underpressure Δu (solid blue/thin line) at 12 m penetration is moderate, just under 300 kPa. Since there is no flow, there is no risk of the soil plug liquefying (the major risks are cavitation if the WD is less than 30 m and steel cylinder implosion if the

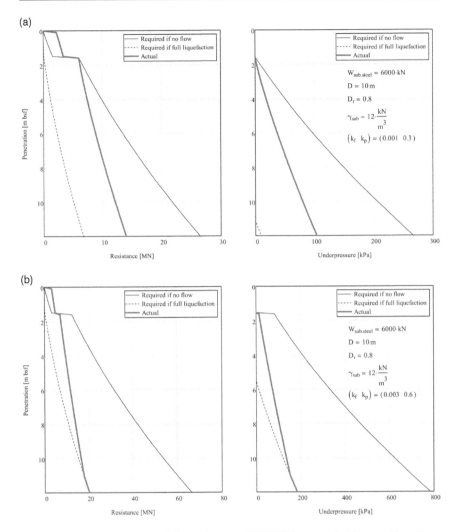

Figure 9.17 Installation in sand. Example using DNV CPT approach. Caisson 10 m diameter. 1.5 m of loose sand (q_c = 2z [MPa,m], Dr ≈ 30%) underlain by dense
sand (q_c = 10 + 1.5z [MPa,m], Dr ≈ 80%). (a) BE k_p and k_f, (b) HE k_p and k_f.

cylinder WT is inadequate). However, seepage significantly decreases penetration resistance R and Δu, and these are given by the red/thick lines. At
12 m penetration, Δu is low, around 100 kPa (was ≈ 300 kPa). The dashed
blue/thin lines are for the extreme case of full liquefaction (i.e. outer skin
friction only – both inner skin friction and wall end-bearing are zero). Unless
full liquefaction occurs (see the High Estimate" bottom of figure), the full
liquefaction lines (plug hydraulic gradient i = i_{crit}) lie below the actual seepage lines (i < i_{crit}). For the Best Estimate k_p and k_f values, the actual seepage
lines never intersect the full liquefaction line.

Figure 9.17b shows the corresponding High Estimate resistance and pressure depth diagrams. The situation changes dramatically: full plug liquefaction first occurs at around 11 m bsf (i.e. L/D $\approx$ 1.1), and the final $\Delta u \approx 175$ kPa (was ≈ 100 kPa) at 12 m penetration. In such cases, where liquefaction occurs only for the last metre of penetration, and only for High Estimate k_p and k_f values, it is tempting to accept the risk and adopt suitable due diligence procedures (Section 9.11.1) during installation.

9.11.3.2 Commentary

Large diameter caissons (D $\approx$ 10 m) are necessary to achieve embedment ratios around L/D $\approx$ 1 to 1.5 in competent sands. This is largely due to the fact that load is proportional to D^2, whereas resistance is proportional only to D.

All profiles are straight lines because CPT q_c varied linearly in each layer. In the real world, greater variability is expected.

The model can be modified for non-uniform caissons, other α_u and i_{sand} relationships. Unlike the clay installation example (Section 9.10.3), retrieval/removal has been omitted, and interested engineers can easily include this. In this case, note that (unlike clay), all Q_{wall} terms practically become zero. Again, $W_{sub,steel}$ counteracts overpressure. Hopefully, friction "set-up" is allowed for by judicious k_f selection.

SWP was 1.5 m for both BE and HE assessments. This coincides with the top surface of the underlying dense sand. Since this depth exceeds (say) 0.5 m, this is satisfactory and is sufficient to start the suction-assisted stage. Had the surface loose sand layer been absent, then SWP $\approx$ 0, and there is a high risk of local piping occurring around the can tip. Special measures (listed in Section 9.9.2) need to be considered.

9.11.4 Friction set-up in sand

Short term (retrieval) set-up is generally not considered in sand but should, however, be considered for removal.

It is generally accepted that skin friction (but not end-bearing) resistance of driven piles in sand increases with (logarithm) time. Mechanisms (see Jardine et al., 2005; Lehane et al., 2005) include

- creep destroying "arching", thereby increasing soil radial effective stress
- aging increasing radial stiffness and/or dilation
- corrosion and/or aging (cementation) causing increase in pile-soil interface friction angle.

Opinions differ as to rate – see Figure 28 of Jardine et al. (2005) and Figure 18 of Schneider et al. (2007).

Additional information is given in:

Jardine et al. (2005)	sand and clay
Lehane et al. (2005)	sand
Jardine et al. (2006)	sand
Schneider et al. (2007)	sand
Karlsrud et al. (2014)	field tests piles sands and clays
Lehane et al. (2014)	centrifuge tests caisson un-aged sands.

With the exception of Lehane et al. (2014), all these studies were for driven pile (not intermediate) foundations.

9.11.5 Back analysis of installation data in sand

Installation in sand is not yet well understood. The number of well-documented field projects and centrifuge tests in the public domain is limited (Panagoulias, 2016). Consequently models are cautious and/or not optimal. Back analysis of field installation data (e.g. Chatzivasileiou, 2014; Frankenmolen et al., 2017) is considered to be an excellent way forward to improve models, reduce the spread between Best Estimate and High Estimate DNV type k_p and k_f coefficients, and hence enable foundations in sand to be more confidently designed for embedment ratios L/D in excess of one.

However, back analysis is challenging – ideally at least three unknowns (typically parameters k_p, k_f and α_u) have to be derived using a single known (penetration resistance R). Also, Section 9.8.4 (CPT Method Coefficient α_u) showed that α_u is a function of foundation penetration, soil plug : soil mass permeability k_f and sand layer thickness. To overcome these complications, back-analyses should adhere to the following rules:

1. Use the same model as for installation prediction.
2. Include the suction assisted zone. This allows for the possibility of altered k_p, k_f values in more competent deeper sands and/or different degrees of disturbance.
3. Keep the α_u model as simple as possible. It is considered reasonable to use the Houlsby and Byrne (2005) α_u – L/D relationship (or similar) for an infinite half space of uniform permeability. This reduces the number of unknowns from three to two.

The challenge (more unknowns than knowns) is dissimilar to that of CPTs – from CPT q_c and R_f it is possible to derive up to two parameters, but it is virtually impossible to derive (say) soil type and strength using only q_c. Research may eventually provide an inexpensive and robust solution to increase the number of knowns – for example, skirt pore pressure measurements. Push-out data ($k_p = 0$) reduces the number of unknowns by one. Hence, specifying (say) 5% extraction and back-analysis may prove beneficial.

9.11.6 Observational method in Peck (1969)

The observational method (Peck, 1969; Eurocode/CEN 2004) may be useful for projects where foundation embedment ratios L/D in excess of one in sand are being considered. Due diligence installation procedures have been given in Section 9.11.1.

Since the observational method is a departure from conventional offshore design practice, details are given. The observational method was described by Peck (1969) and forms part of Eurocode geotechnical design (CEN, 2004).

Peck (1969) notes:

In brief, the complete application of the (observational) method embodies the following ingredients.

(a) Exploration sufficient to establish at least the general nature, pattern and properties of the deposits, but not necessarily in detail.

(b) Assessment of the most probable conditions and the most unfavourable conceivable deviations from these conditions. In this assessment, geology often plays a major role.

(c) Establishment of the design based on a working hypothesis of behaviour anticipated under the most probable conditions.

(d) Selection of quantities to be observed as construction proceeds and calculation of their anticipated values on the basis of the working hypothesis.

(e) Calculation of values of the same quantities under the most unfavourable conditions compatible with the available data concerning the subsurface conditions.

(f) Selection in advance of a course of load or modification of design for every foreseeable significant deviation of the observational findings from those predicted on the basis of the working hypothesis.

(g) Measurement of quantities to be observed and evaluation of actual conditions.

(h) Modification of design to suit actual conditions.

CEN (2004) proposes the following (Section 2.4, 4P):

If no reliable calculation model is available for a specific limit state, analysis of another limit state shall be carried out using factors to ensure that exceeding the specific limit state considered is sufficiently improbable. Alternatively, design by prescriptive measures, experimental models and load tests, or the observational method, shall be performed.

9.11.7 Observational method in CEN (2004)

1. When prediction of geotechnical behaviour is difficult, it can be appropriate to apply the approach known as "the observational method", in which the design is reviewed during construction.

2. The following requirements shall be met before construction is started:
   - acceptable limits of behaviour shall be established;
   - the range of possible behaviour shall be assessed and it shall be shown that there is an acceptable probability that the actual behaviour will be within the acceptable limits;
   - a plan of monitoring shall be devised, which will reveal whether the actual behaviour lies within the acceptable limits. The monitoring shall make this clear at a sufficiently early stage, and with sufficiently short intervals to allow contingency actions to be undertaken successfully;
   - the response time of the instruments and the procedures for analysing the results shall be sufficiently rapid in relation to the possible evolution of the system;
   - a plan of contingency actions shall be devised, which may be adopted if the monitoring reveals behaviour outside acceptable limits.

3. During construction, the monitoring shall be carried out as planned.

4. The results of the monitoring shall be assessed at appropriate stages and the planned contingency actions shall be put into operation if the limits of behaviour are exceeded.

5. Monitoring equipment shall either be replaced or extended if it fails to supply reliable data of appropriate type or in sufficient quantity.

9.11.8 General

Installation studies are difficult and opinions/models/results differ widely. Key challenges usually include:

(a) unit inner skin friction and end-bearing reduction in the sand aquifer due to upwards seepage in soil plug

(b) water pocket in soil plug at the clay/sand interface. This is because installation stops if separation occurs and the clay plug meets the foundation top plate (see Figure 8.2d)

(c) hydraulic fracture of the clay plug (see Figure 8.2f). This is perceived to be generally associated with competent clay and moderate underpressures.

Background reading includes:

- Tran (2005)   sand and silt layers
- Watson et al. (2006) clay over sand, clay plug heave
- Senders et al. (2007) clay over sand and clay plug hydraulic fracture model
- Romp (2013)   clay over sand, clay plug heave and hydraulic fracture (cracking)
- Plug heave   see Section 9.10.2 (on plug heave in clay) and Section 9.11.2 (on models for sand).

9.11.9 Water pocket model

This section is about a model for water pocket development (until now unpublished). This is an improved version of the original (1996) Fugro concept but has been greatly improved for water flow. It was used on a project considering 15 m diameter caissons as FPSO anchors under 24 MN lateral load.

The following questions arise when installing suction caissons in mixed clay and sand layers: (a) Is there a possibility of clay plug–sand plug separation, causing a water pocket to be formed at the sand–clay interface? (b) Is there any reduction in inner friction and tip resistance in the sand layers? The model is sketched in Figure 9.18 and is analogous to a piston (the

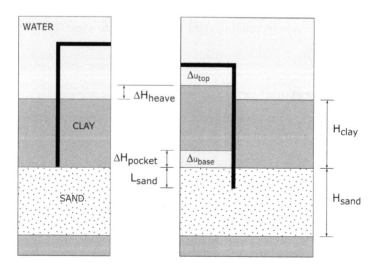

Figure 9.18 Water pocket. Water pocket, height ΔH_{pocket}, at bottom of clay plug. Clay plug height increase $\Delta H_{heave} = \Delta H_{pocket}$. Foundation penetration L_{sand} into sand layer.

impermeable clay plug) being pulled out of a cylinder embedded in permeable material (the sand aquifer). The soil profile consists of clay (height H_{clay}) underlain by sand (H_{sand}). The caisson is penetrating L_{sand} into the sand aquifer. If there is fluid flow Q_{flow} from the sand, a water pocket, height ΔH_{pocket}, will form with an underpressure Δu_{base}. The upwards-acting differential force, magnitude $(\Delta u_{top} - \Delta u_{base}) A_{plug}$, pushes the clay plug upwards. There will be no water pocket if Q_{flow} is zero. Conversely, according to the model, a water pocket will start to form as soon as $Q_{flow} > 0$.

Model assumptions include the following:

(a) The clay plug is impermeable. Hence, there is no fluid flow (i.e. no head drop) within the clay plug
(b) The sand response is fully drained. Hence there are no excess pore pressures (effective stress = total stress, pore pressures are steady state).

The objective is to find the sand aquifer flowrate Q_{flow} at any penetration L_{sand}. From this, and knowing the time to penetrate the sand aquifer, ΔH_{pocket}, the required water pocket height, i.e. clay plug heave, ΔH_{heave}, can be computed.

The two steady state flow approximate equations are:

(a) sand annulus + plug : flowrate $Q_{flow} - \Delta u_{base}$

$$Q_{flow} \approx \left[8\pi / \left(13\gamma_{water}\right)\right]\left[\left(k_{sand}H_{sand}\right) / \alpha_{HLD} \Delta u_{base}\right] \tag{9.65}$$

sand annulus + plug: upwards hydraulic gradient $i_{plug} - \Delta u_{base}$

$$i_{plug} \approx \left\{\left[1-\left(1/\alpha_{HLD}\right)\right]/\left[L_{sand}\gamma_{water}\right]\right\}\Delta u_{base} \tag{9.66}$$

There are also two equilibrium equations and another two linking sand resistances to the hydraulic gradient in the sand plug:

(b) forces on caisson steel

$$\Delta u_{top}A_{plug} + W_{sub,steel} = F_{i,clay} + F_{o,clay} + F_{i,sand} + F_{o,sand} + Q_{w,sand} \tag{9.67}$$

forces on clay plug

$$\Delta u_{top}A_{plug} = \Delta u_{base} + \left(F_{i,clay} + W_{sub\_clay}\right) \tag{9.68}$$

friction and wall end-bearing resistances in sand

$$F_{i,sand} = F_{i,sand,no\_seep}\left(1 - i_{base}\right) \tag{9.69}$$

$$Q_{w,sand} = Q_{w,sand,no\_seep}\left(1 - i_{base}\right) \tag{9.70}$$

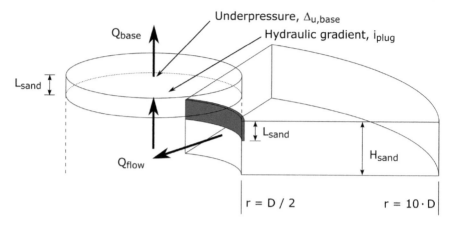

Figure 9.19 Water pocket. Steady state radial fluid flow – sand plug and aquifer annulus components are connected with no change in hydraulic head.

Assumptions, limitations and reasonableness of these equations include:

Ad (a) and (b): as sketched in Figure 9.19, the sand layer has been sub-divided into two components, an annulus (height H_{sand}, inner and outer radii D/2 and 10 D, with a cut-off height L_{sand} equal to the tip penetration) representing the outer portion of the confined sand aquifer, and the sand plug (diameter D, height L_{sand}). The volume of sand in the aquifer directly below the sand plug is missing, for which it is assumed that a frictionless pipe transmits all radial flow out of the aquifer annulus into axial upwards flow in the sand plug. Steady state axisymmetric fluid flow FEA were made of the annulus for varying L_{sand}/H_{sand} ratios (0–1) to obtain $Q_{flow} – \Delta u_{base}$ and $Q_{flow} – i_{plug}$ relationships. Both are considered reasonable approximations and are generally cautious (i.e. they overpredict Q_{flow} and ΔH_{pocket}), as shown in Figure 9.20. Better (slightly more accurate) project specific relationships are possible (e.g. partially penetrating well theory or FEA), but, as will be shown, heave assessments are strongly dependent on k_{sand}.

These considerations suggest that there cannot be installation resistance reduction in sand without a water pocket – there must be flow. No-flow resistances apply if there is no seepage.

Ad (c): force equilibrium - caisson. The caisson is penetrating into the soil; hence the downwards-acting underpressure and caisson weight are opposed by upwards acting soil resistances. Note that the clay resistances remain constant, but in the sand they are a function of tip penetration L_{sand} and i_{plug}. In addition (unlike free-fall penetration), the buoyancy force is in water, not soil penetration dependent, and therefore constant.

Ad (d): force equilibrium – clay plug. Again, caisson penetration is assumed. Hence, inner friction resistance $F_{i,clay}$ and underpressure $\Delta u_{base} A_{plug}$ both act downwards opposing caisson motion.

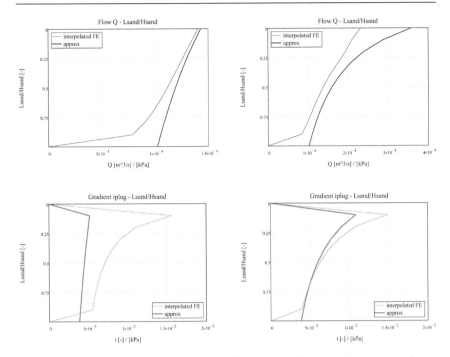

Figure 9.20 Water pocket. Comparisons Q_{flow} versus Δu_{base} and i_{plug} versus Δu_{base} approximate relationships with axisymmetric FEA at $H_{sand}/D = 0.4$ and 1.0. Diameter $D = 18.5$ m, $k_{sand} = 1e-4$ m/s.

Ad (e): caisson friction and end-bearing in sand. Opinions differ, and a simple linear reduction of both inner friction and tip resistance (and no outer friction resistance reduction) with sand seepage gradient i_{plug} has been assumed. Since inner friction and end-bearing both become zero when $i_{plug} = 1$, i_{plug} cannot exceed unity. Other end-bearing possibilities include 50% when $i_{plug} = 1$, but ΔH_{pocket} is more sensitive to $Q_{suction}$ and k_{sand} than resistance.

There are six variables and six equations. Hence the five unknowns (i_{base}, Δu_{top}, Δu_{base}, $F_{i,sand}$ and $Q_{w,sand}$), all of which are L_{sand} dependent, can be derived for a given Q_{flow} using the equations:

$$Q_{flow} = \left(8\pi/13\right)\left(k_{sand}H_{sand}L_{sand} / \alpha_{HLD}\right) \times$$
$$\left[\left(F_{o,clay} + F_{i,sand,no\_seep} + F_{o,sand,no\_seep} + Q_{w,sand,no\_seep}\right) - \left(W_{sub,steel} + W_{sub,clay}\right)\right] /$$
$$\left[1 - \left(1/\alpha_{HLD}\right)\right]\left(F_{i,sand,no\_seep} + Q_{w,sand,no\_seep}\right) + A_{plug}L_{sand}\gamma_{water}\right] \tag{9.71}$$

$$\Delta u_{base} = Q_{flow} / \left\{\left[8\pi/\left(13\gamma_{water}\right)\right] / \left[k_{sand}H_{sand} / \alpha_{HLD}\right]\right\} \tag{9.72}$$

$$i_{plug} = \min\left(\left\{\left[1 - \left(1/\alpha_{HLD}\right)\right] / \left[L_{sand}\gamma_{water}\right]\right\}\Delta u_{base}, 1\right) \tag{9.73}$$

$$\Delta u_{top} = \Delta u_{base} + \left(F_{i,clay} + W_{sub,clay}\right) / A_{plug} \tag{9.74}$$

$$F_{i,sand} = F_{i,sand,no\_seep} \left(1 - i_{plug}\right) \tag{9.75}$$

$$Q_{w,sand} = Q_{w,sand,no\_seep} \left(1 - i_{plug}\right) \tag{9.76}$$

If the aquifer flow rate Q_{flow} is negative, then there is no water pocket, and the no-seepage values should be used, namely:

$$Q_{flow} = 0 \tag{9.77}$$

$$\Delta u_{base} = 0 \tag{9.78}$$

$$i_{plug} = 0 \tag{9.79}$$

$$\Delta u_{top} = \left[\left(F_{i,clay} + F_{o,clay} + F_{i,sand,no\_seep} + F_{o,sand,no\_seep} + Q_{w,sand,no\_seep}\right) - W_{sub,steel}\right] / A_{plug} \tag{9.80}$$

$$F_{i,sand} = F_{i,sand,no\_seep} \tag{9.81}$$

$$Q_{w,sand} = Q_{w,sand,no\_seep} \tag{9.82}$$

where, noting that subscript "no_seep" means "with suction assistance but no seepage/fluid flow":

α_{HLD}	$= (1 + 2.5\ H_{sand}\ L_{sand}/D^2)$
ΔH_{pocket}	= water pocket height (= clay plug heave ΔH_{heave})
Δu_{top}	= water (under) pressure at clay plug top
Δu_{base}	= water (under) pressure in water pocket
$\gamma_{sub,clay}$	= clay plug submerged unit weight
γ_{water}	= (sea) water unit weight
A_{plug}	= soil plug area
D	= caisson diameter
$F_{i,clay}$	= caisson inner wall friction – clay
$F_{o,clay}$	= caisson outer wall friction – clay
$F_{i,sand}$	= caisson inner wall friction – sand, fluid flow
$F_{i,sand,no\_seep}$	= caisson inner wall friction – sand, no seepage
$F_{o,sand}$	= caisson outer wall friction – sand, fluid flow

$F_{o,sand,no\_seep}$ = caisson outer wall friction – sand, no seepage

H_{clay} = clay layer and clay plug thickness

H_{sand} = sand aquifer thickness

i_{plug} = sand plug (vertical) average seepage gradient

$k_{f,clay}, k_{f,sand}$ = DNV empirical coefficients – unit skin friction resistance in clay and sand

$k_{p,clay}, k_{p,sand}$ = DNV empirical coefficients – unit end-bearing resistance in clay and sand

k_{sand} = sand aquifer permeability

L_{sand} = caisson tip penetration into sand aquifer

N_k = empirical CPT cone factor

N_{Lsand} = number of penetration depths L_{sand} (= 0 through H_{sand})

$pen_{rate,sand,avg}$ = caisson average penetration rate into sand aquifer

Q_{base} = sand plug flowrate (into water pocket)

Q_{flow} = sand aquifer flowrate (into base of sand plug)

$Q_{flow,avg}$ = average Q_{flow} during aquifer penetration

$Q_{suction}$ = suction pump flowrate

$Q_{w,sand}$ = caisson wall tip resistance – sand, fluid flow

$Q_{w,sand,no\_seep}$ = caisson wall tip resistance – sand, no seepage

R_{no_seep} = soil resistance without suction assistance

$R_{suction}$ = soil resistance with suction assistance

$t_{pen,Lsand}$ = time taken to penetrate sand aquifer to L_{sand}

V_{water} = water plug volume at time $t_{pen,Lsand}$

$W_{sub,steel}$ = caisson submerged weight (including ballast)

$W_{sub,clay}$ = clay plug submerged unit weight = $\gamma_{sub,clay} A_{plug} H_{clay}$

Steps to find ΔH_{pocket} are as follows:

1. Perform steady state axisymmetric fluid flow sums to obtain Q_{flow} – Δu_{base} and Q_{flow} – i_{plug} relationships for varying L_{sand}/H_{sand} ratios (0–1); see Figure 9.20. Use these relationships for the first two equations in step 2 following.
2. Set up six equations and solve for 6 unknowns (see pseudo code that follows).
3. Repeat step 2 for selected L_{sand} values to obtain Q_{flow} variation.
4. Hence, obtain average flowrate $Q_{flow,avg}$ whilst the caisson penetrates the sand aquifer.
5. Knowing the average caisson penetration rate $pen_{rate,sand,avg}$ and A_{plug}, values of V_{water} (water pocket volume), the required ΔH_{pocket} value can be computed. See the water pocket model example that follows.

The corresponding pseudo code for the water pocket model is:

```
program water_pocket
initialise and evaluate constants; assume DNV CPT qc model
   for resistances
```
$F_{*,clay}$: skin friction resistances by integrating q_c from 0 to
   H_{clay}
$F_{i,clay} = \pi\ D_i\ k_{f,clay} \int q_{c,z}\ dz$ and $F_{o,clay} = \pi\ D\ k_{f,clay} \int q_{c,z}\ dz$
$pen_{rate,sand,avg}$: caisson average penetration rate
$A_{plug} = \pi\ D_i^2/4$
$pen_{rate,sand,avg} = Q_{suction}\ /A_{plug}$
$W_{sub,clay} = A_{plug}\ H_{clay}\ \gamma_{sub,clay}$
$N_{Lsand} = 21$
```
zero matrix ANS (size NLsand rows by 10 + 4 = 14 columns)
k = 0
loop tip penetration Lsand from 0 to Hsand in k equal steps:
   k = k + 1
```
   $L_{sand} = max(L_{sand}, 0.001\ m)$
   $F_{I,sand,no\_seep}$ and $F_{I,sand,no\_seep}$: skin friction resistances by
```
      integrating qc from Hclay to Hclay + Lsand
```
   $Q_{w,sand,no\_seep}$: end-bearing resistance using q_c at depth $(H_{clay}$
      $+ L_{sand})$
   $F_{i,sand,no\_seep} = \pi\ D_i\ k_{f,sand} \int q_{c,z}\ dz$ and $F_{o,sand,no\_seep} = \pi\ D\ k_{f,sand}$
      $\int q_{c,z}\ dz$
   $Q_{w,sand,no\_seep} = A_{wall}\ k_{p,sand}\ q_{c,z}$
   $\alpha_{HLD} = (1 + 2.5\ H_{sand}\ L_{sand}/D^2)$
   $Q_{flow}, \Delta u_{base}, i_{plug}, F_{i,sand}, Q_{w,sand}$ and Δu_{top} using Equations (9.71)
```
      through (9.76)
```
   $R_{no\_seep} = F_{i,clay} + F_{o,clay} + F_{i,sand,no\_seep} + F_{o,sand,no\_seep} +$
   $Q_{w,sand,no\_seep}$
   $F_{o,sand} = F_{o,sand,no\_seep}$
   $R_{suction} = F_{i,clay} + F_{o,clay} + F_{i,sand} + F_{o,sand} + Q_{w,sand}$
```
   store in kth row of matrix ANS:
```
      $L_{sand}, L_{sand}\ /H_{sand}, R_{no\_seep}$ and $R_{suction}$ in columns 1 through 4
      $Q_{flow}, i_{plug}, \Delta u_{top}$ and Δu_{base} in columns 5, 6, 9 and 10
      $F_{i,sand}$ and $Q_{w,sand}$ in columns 7 and 8
```
end loop on tip penetration Lsand
k = 0
loop tip penetration Lsand from 0 to Hsand in k equal steps:
   k = k + 1
```
   $Q_{flow,avg} = \sum Q_{flow}$ (i = 1 to k)
   $Q_{flow,avg} = Q_{flow,avg}\ /k$
   $L_{sand} = ANS(k,1)$
   $t_{pen,Lsand} = L_{sand}/pen_{rate,sand,avg}$
```
```

$$V_{water} = t_{pen,Lsand} \, Q_{flow,avg}$$

$$\Delta H_{pocket} = V_{water}/A_{plug}$$

store in k^{th} row of matrix ANS:

$Q_{flow,avg}$, $t_{pen,Lsand}$, V_{water}, ΔH_{pocket} in columns 11-14

end loop on tip penetration L_{sand}

exit with matrix ANS

matrix ANS (N_{Lsand} rows, 14 columns) contents:

col# 01: L_{sand}	caisson tip penetration into sand aquifer	[m]
col# 02: L_{sand} / H_{sand}	penetration/aquifer height ratio	[-]
col# 03: R_{no_seep}	soil resistance without suction assistance	[MN]
col# 04: $R_{suction}$	soil resistance with suction assistance	[MN]
col# 05: Q_{flow}	sand aquifer flowrate (into base of sand plug)	[m³/s]
col# 06: i_{plug}	sand plug (vertical) average seepage gradient	[-]
col# 07: $F_{i,sand}$	caisson inner wall friction – sand, suction	[MN]
col# 08: $Q_{w,sand}$	caisson wall tip resistance – sand, suction	[MN]
col# 09: $\Delta_{u,top}$	water (under) pressure at clay plug top	[kPa]
col# 10: $\Delta_{u,base}$	water (under) pressure in water pocket	[kPa]
col# 11: $Q_{flow,avg}$	average flowrate Q whilst in aquifer at L_{sand}	[m³/s]
col# 12: $t_{pen,Lsand}$	time taken to penetrate sand aquifer to L_{sand}	[s]
col# 13: V_{water}	water pocket volume at time $t_{pen,Lsand}$	[m³]
col# 14: ΔH_{pocket}	water pocket height (= clay plug heave ΔH_{heave})	[m]

Plot col#14 (x, ΔH_{pocket}) versus col# 01 (y, L_{sand}) *water pocket height (= clay plug heave) – penetration.*

9.11.9.1 Example – water pocket model

The following example is for competent soil. DNV CPT High Estimate $k_{p,sand}$, $k_{f,sand}$ sand data, a high sand permeability k, and a typical suction pump flowrate $Q_{suction}$ (sufficient to obtain ≈ 1 mm/s penetration rate) are used to assess high water pocket height values. Input data for a typical North Sea type problem (basically 4 m of 70 kPa clay, underlain by sand q_c ≈ 20 MPa) are:

Pile: D = 10 m, WT = D/200 = 50 mm, $W_{sub,steel}$ = 6000 kN
        (D_i = 9.90 m, $A_{s,i}$ = π D_i = 31.10 m, $A_{s,o}$ = π D = 31.41 m, A_p = A_{wall}
        = 1.565 m², A_{plug} = 77.0 m²)

Soil: clay H_{clay} = 4 m, $q_{c,clay}$ = 1 + 0.1 z [MPa, m], γ_{sub} = 9 kN/m³
        sand H_{sand} = 6 m, $q_{c,sand}$= 20 + 0.5 (z - H_{clay}) [MPa, m], k_{sand} = 1e-4 m/s

Pile – Soil – Fluid:
        γ_{water} = 10 kN/m³
        $k_{p,clay}$, $k_{f,clay}$ = 0.6, 0.05 and $k_{p,sand}$, $k_{f,sand}$ = 0.6, 0.003
        $Q_{suction}$ = 83.3e-3 m³/s (300 m³/hour)

Sand permeability can range between k_{sand} = 1e-4 m/s and k_{sand} = 1e-6 m/s; HE taken.

The sand layer has been analysed at n_{Lsand} = 21 equally spaced depth intervals.

Pseudo code constants are:

$F_{i,clay}$ = 7.46 MN and $F_{o,clay}$ = 7.54 MN
$W_{sub,clay}$ = 2.77 MN
$pen_{rate,sand,avg}$ = 1.08 mm/s

Halfway through the sand layer, at L_{sand} = 3 m, we have

$F_{i,sand,no\_seep}$ = 5.81 MN and $F_{o,sand,no\_seep}$ = 5.87 MN
$Q_{w,sand,no\_seep}$ = 20.2 MN
α_{HLD} = 1.45
Q_{flow} = 7.08e-3 m³/s   Equation (9.71)
Δu_{base} = 88.6 kPa   Equation (9.72)
i_{plug} = 0.916   Equation (9.66)
Δu_{top} = 222 kPa   Equation (9.68)
$F_{i,sand}$ = 0.488 MN   Equation (9.69)
$Q_{w,sand}$ = 1.69 MN   Equation (9.70)

and

$F_{o,sand}$ = 5.87 MN
R_{no_seep} = $F_{i,clay}$ + $F_{o,clay}$ + $F_{i,sand,no\_seep}$ + $F_{o,sand,no\_seep}$ + $Q_{w,sand,no\_seep}$ = 46.8 MN
$R_{suction}$ = $F_{i,clay}$ + $F_{o,clay}$ + $F_{i,sand}$ + $F_{o,sand}$ + $Q_{w,sand}$ = 23.1 MN
$Q_{flow,avg}$ = 6.43e-3 m³/s
$t_{pen,Lsand}$ = 2270 s
V_{water} = 17.95 m³
ΔH_{pocket} = 0.23 m

Figure 9.21 graphs the results. Because of the model complexity, the basic components are also plotted to gain confidence before discussing the key plot (ΔH_{pocket} versus L_{sand}) on Figure 9.21b.

The two right hand underpressure versus penetration plots on Figure 9.21a show that:

- self-weight penetration is just under 2 m into the 70 kPa clay. This is sufficient to start the suction assisted penetration stage.
- underpressure at the base of the clay is just over 100 kPa.
- within the underlying sand, a water pocket is created at the start of penetration, and this is the reason why the blue line (with pocket) plots below the red/thick line (no pocket). Underpressures increase by

around 100 kPa to just under 200 kPa when the tip penetrates the sand, increasing to 222 kPa halfway through (L_{sand} = 3 m) and around 260 kPa at the base of the sand aquifer (L_{sand} = 6 m). These magnitudes are normally reasonable for structural engineers. The red/thick lines denote results had no water pocket been formed. In this case, underpressures are significantly higher (around 350, 500 and 700 kPa respectively), for which shell buckling and/or top plate design would be of concern.

On Figure 9.21b, it can be seen that:

- aquifer flow rates, just over 20 m³/h for the first 3 m of sand penetration, are creating the water pocket. They are less than 10% of $Q_{suction}$ (300 m³/h).

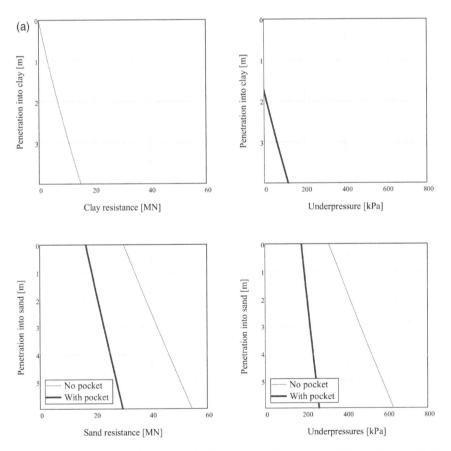

Figure 9.21 Water pocket example. Foundation D = 10 m, WT = 40 mm. Soil profile: 4 m of 100 kPa clay over 20 MPa sand, k = 1e-4m/s, $Q_{suction}$ = 250 m³/hour: (a) soil resistances and underpressures versus penetration.

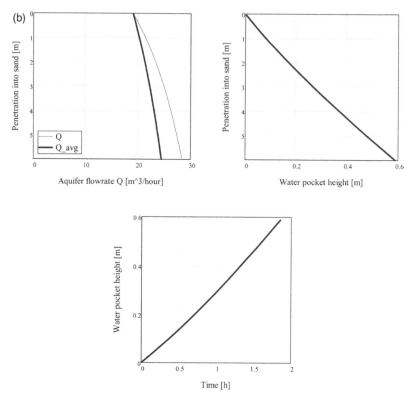

Figure 9.21 Continued: Water pocket example. Foundation D = 10 m, WT = 40 mm. Soil profile: 4 m of 100 kPa clay over 20 MPa sand, k = 1e-4m/s, $Q_{suction}$ = 250 m³/hour: (b) water pocket flowrate and height versus penetration and time.

- a water pocket is created immediately, at the start of penetration into the sand. The water pocket height ΔH_{pocket} increases almost linearly with penetration: at 3 m penetration, ΔH_{pocket} = 0.232 m and the final ΔH_{pocket} = 0.503 m.
- applying $Q_{suction}$ (300 m³/h), 3 m and 6 m penetration into the sand take ≈ 0.77 and 1.53 hours respectively.

Finally, the four plots on Figure 9.21c show that

- because there is a water pocket, i_{plug}, the sand plug seepage gradient is non-zero and increases with increasing penetration. This is caused by increasing underpressure. A critical state "liquefaction" (i_{plug} = 1) is present from just over 5 m penetration into the sand.
- both inner friction and wall tip components decrease due to seepage, and are zero below 5 m.
- the differential water pressure (Δu_{top} - Δu_{base}) is ≈ 135 kPa and constant. This is as expected – when (Δu_{top} – Δu_{base}) A_{plug} is the force moving the clay plug upwards.

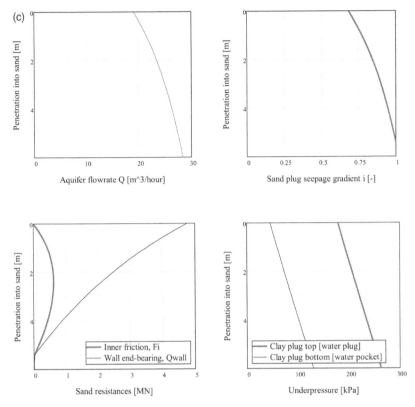

Figure 9.21 Continued: Water pocket example. Foundation D = 10 m, WT = 40 mm. Soil profile: 4 m of 100 kPa clay over 20 MPa sand, k = 1e-4m/s, $Q_{suction}$ = 250 m^3/hour:(c) water pocket flowrate, sand plug gradient, sand resistances and underpressures versus penetration.

Geotechnical engineers should make sensitivity analyses to familiarise themselves with complex models. By changing original parameters one by one, their effect on ΔH_{pocket} could be found, and results are given in Table 9.9.

It is seen that ΔH_{pocket} (which is the same as ΔH_{heave}) is very sensitive to $Q_{suction}$ and k_{sand}. ΔH_{pocket} is inversely proportional to $Q_{suction}$ and directly proportional to k_{sand}. Minimum ΔH_{pocket} occurs when caisson penetration rate is maximised (i.e. get through the sand as fast as possible). This implies that $Q_{suction}$ should be as large as possible in "clay over sand" soil profiles. These ΔH_{pocket} values were at 6 m penetration into the sand, i.e. a caisson embedded length L = 10 m (L/D = 1.1). In practice, because of possible liquefaction from $\approx$ 4 m into the sand, a lower L value, probably around 8 m (giving L/D = 0.8) would be studied in more detail. This would also give less onerous clay plug heave ($\approx$ 0.3 m) and underpressures ($\approx$ 200 kPa).

Table 9.9 also shows that ΔH_{pocket} is moderately sensitive to caisson geometry (D < 7.5 m are unlikely to be effective) and to the sand layer thickness H_{sand} (aquifer inflow Q_{flow} is proportional to transmissivity T = $k_{sand} H_{sand}$).

Table 9.9 Water Pocket Model – Sensitivity Analyses.

Parameter(s)	Original and revised parameter value(s)			Revised ΔH_{pocket} [m]	Revised/ original ΔH_{pocket} [–]	Remarks
$Q_{suction}$	300	600	[m³/h]	0.251	0.5	Original: 1 mm/s penetration rate
k_{sand}	1e-4	1e-5	[m/s]	0.05	0.1	Original: HE
k_{sand}	1e-4	1e-6	[m/s]	5.1e-3	0.01	Original: HE
$k_{p,clay}$	0.6	0.4	[–]	0.273	0.543	Original: HE
$k_{f,clay}$	0.05	0.03	[–]			Revised: BE
$k_{p,sand}$	0.6	0.6	[–]			
$k_{f,sand}$	0.003	0.001	[–]			
D	10	15	[m]	0.791	1.573	WT = D/200, $W_{sub,steel} \propto D^2$
WT	50	75	[mm]			no liquefaction in sand
$W_{sub,steel}$	6000	13500	[kN]			
D	10	7.5	[m]	0.345	0.686	WT = D/200, $W_{sub,steel} \propto D^2$
WT	50	37.5	[mm]			liquefaction from 1 m in sand
$W_{sub,steel}$	6000	3375	[kN]			
H_{clay}	4	6	[m]	0.560	1.113	liquefaction from 2.5 m in sand
H_{clay}	4	3	[m]	0.478	0.951	no liquefaction in sand
H_{sand}	6	9	[m]	0.838	1.665	liquefaction from 4 m in sand
H_{sand}	6	3	[m]	0.197	0.391	no liquefaction in sand

Notes:
BE: Best Estimate, HE: High Estimate
Water pocket height ΔH_{pocket} = clay plug heave ΔH_{heave}
Original ΔH_{pocket} = 0.503 m, H_{sand} = 6 m, i.e. $\Delta H_{pocket}/H_{sand}$ = 8%
ΔH_{pocket} is $\approx$ proportional to tip penetration in sand; H_{pocket} values are when tip reaches the base of the sand layer.

9.11.10 Reverse punch-through failure

The reverse punch-through failure mode is illustrated in Figure 9.22 for a caisson being installed into clay over sand. Unlike the water pocket model, the clay layer extends below the caisson tip. There is a possibility of reverse punch-through (uplift of the complete clay plug) occurring, even though the caisson tip level is within the clay. Conventional punch-through is studied in detail for "clay over sand" profiles below spudcans (ISO 19905-1:2016 and Figure 2.1a).

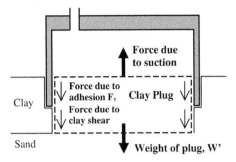

Figure 9.22 Reverse punch-through failure model: Forces acting on the clay plug. No reverse-end-bearing (Cotter, 2009).

The model assumes zero reverse end-bearing (REB), and no water pocket development, at the clay/sand interface. There are two equilibrium equations: forces on the caisson steel (during installation) and forces on the complete clay plug (reverse punch-through). Note that $W_{sub,steel}$ assists penetration but $W_{sub,clay,RPT}$ opposes reverse punch-through. From these equations, the corresponding underpressures Δu can be computed. Reverse punch-through failure is possible when $\Delta u_{RPT} > \Delta u_{install}$. The four equations are:

$$\Delta u_{install} A_{plug} + W_{sub,steel} = F_{i,clay,adhesion} + F_{o,clay,adhesion} + Q_{w,clay} \qquad (9.83)$$

$$\Delta u_{RPT} A_{plug} = F_{i,clay,adhesion} + F_{i,clay,shear} + W_{sub,clay,RPT} \qquad (9.84)$$

$$\Delta u_{install} = \left[\left(F_{i,clay,adhesion} + F_{o,clay,adhesion} + Q_{w,clay} \right) - W_{sub,steel} \right] / A_{plug} \quad (9.85)$$

$$\Delta u_{RPT} = \left[\left(F_{i,clay,adhesion} + F_{i,clay,shear} \right) + W_{sub,clay,RPT} \right] / A_{plug} \qquad (9.86)$$

where

σ_{vo}	= total soil vertical stress at depth z = $(\gamma_{sub,clay} + \gamma_{water})$ z
$\gamma_{sub,clay}$	= clay submerged unit weight
γ_{water}	= (sea) water unit weight
A_{plug}	= soil plug area
A_p	= caisson wall tip area
$A_{s,i}$	= caisson inner perimeter
$A_{s,o}$	= caisson outer perimeter
$F_{i,clay,adhesion}$	= caisson inner wall friction – clay adhesion above tip
$F_{o,clay,adhesion}$	= caisson outer wall friction – clay adhesion above tip
$F_{i,clay,shear}$	= caisson inner wall friction – clay shear below tip
H_{clay}	= clay layer and clay plug thickness
$k_{f,clay}$	= DNV empirical coefficient – unit skin friction resistance in clay
$k_{p,clay}$	= DNV empirical coefficient – unit end-bearing resistance in clay

L_{clay} = caisson tip penetration into clay layer
N_k = empirical CPT cone factor
$Q_{w,clay}$ = caisson wall tip resistance – clay
$q_{c,clay,z}$ = CPT cone tip resistance at depth z
$R_{install}$ = soil penetration resistance
R_{RPT} = soil reverse punch-through resistance
$s_{u,z}$ = clay equivalent undrained soil shear strength at depth z
$W_{sub,clay,RPT}$ = punched through clay plug submerged weight =
                     $\gamma_{sub,clay}\, A_{plug}\, H_{clay}$
$W_{sub,steel}$ = caisson submerged weight (including ballast)
$\Delta u_{install}$ = under pressure for caisson installation
Δu_{RPT} = under pressure causing reverse punch-through

The corresponding yield function can be derived from Cotter Equation 2.6.12 and is

$$yf_{RPT} = Y\,grad_1 + intercept_1 - X \qquad (9.87)$$

where

$grad_1$ $= \alpha\,\pi\,(H_{clay}/D) + \pi\,N_{c,strip}\,(WT/D)$
$intercept_1$ $= (H_{clay}/D)\,(\pi\,WT/D - \pi/4)$
X $= W_{sub,steel}/(\gamma_{sub,clay}\,D^3)$
Y $= s_{u,av}/(\gamma_{sub,clay}\,D)$
$s_{u,av,Hclay}$ = average s_u in clay layer
$N_{c,strip}$ = bearing capacity factor at caisson tip

Due to assuming $\Delta u_{install}$ acts over $\pi\,D^2/4$, not $\pi\,D_i^2/4$, equation 9.12.19 is slightly approximate. Reverse punch-through will not occur if $yf_{RPT} < 0$.

9.11.10.1 Example – reverse punch-through

The following example is for competent soil. DNV CPT q_c data are used to assess underpressures $\Delta u_{install}$ and Δu_{RPT}. The input data are consistent with those used for the water pocket example. The data are for a typical North Sea type problem (basically 4 m of 70 kPa clay over sand) and are:

Soil: clay H_{clay} = 4 m, $q_{c,clay,z}$ = 1 + 0.1 z [MPa, m], γ = 19 kN/m^3,
       γ_{sub} = 9 kN/m^3, N_k = 17
       (equivalent $s_{u,z}$ = ($q_{c,clay,z} - \sigma_{vo}$)/$N_k$ ≈ 60 + 5 z [kPa, m])
Pile: D = 10 m, WT = 50 mm, $W_{sub,steel}$ = 6000 kN
       (D_i = 9.9 m, $A_{s,i}$ = $\pi\,D_i$ = 31.1 m, $A_{s,o}$ = $\pi\,D$ = 31.4 m, A_p = 1.563 m^2,
       A_{plug} = 76.977 m^2)
Pile –Soil:
       γ_{water} = 10 kN/m^3

$k_{p,clay}$, $k_{f,clay}$ = 0.6, 0.05 (High Estimates)

The calculations to find and compare underpressures $\Delta u_{install}$ and Δu_{RPT} are straightforward. At L_{clay} = 2 m (half way through the clay), we have:

$q_{c,clay,z}$ = 1000 + 0.1 L_{clay} = 1200 kPa (at 1 m depth)
$F_{i,clay,adhesion}$ = $A_{s,I}$ L_{clay} $k_{f,clay}$ $q_{c,clay,z}$ = 3421 kN
$F_{o,clay,adhesion}$ = $A_{s,o}$ L_{clay} $k_{f,clay}$ $q_{c,clay,z}$ = 3456 kN
$S_{u,z}$ = $(q_{\cdot c,clay,z} - \sigma_{vo}) / N_k$ = 73.1 kPa (at {H_{clay} – L_{clay}} = 3 m depth)
$F_{i,clay,shear}$ = $A_{s,I}$ (H_{clay} – L_{clay}) $s_{u,z}$ = 4548 kN
$q_{c,clay,z}$ = 1000 + 0.1 L_{clay} = 1200 kPa (at 2 m depth)
$Q_{w,clay}$ = A_p $k_{p,clay}$ $q_{c,clay,z}$ = 1125 kN
$W_{sub,clay,RPT}$ = γ_{sub} A_{plug} H_{clay} = 2771 kN
$R_{install}$ = $F_{i,clay,adhesion}$+ $F_{o,clay,adhesion}$ + $Q_{w,clay}$ = 8002 kN
R_{RPT} = $F_{i,clay,adhesion}$+ $F_{i,clay,shear}$ = 7969 kN
$\Delta u_{install}$ = $(R_{install} - W_{sub,steel}) / A_{plug}$ = 26.0 kPa
Δu_{RPT} = $(R_{RPT} + W_{sub,clay,RPT}) / A_{plug}$ = 139 kPa

Since $\Delta u_{install}$ < Δu_{RPT}, reverse punch-through is not occurring at 2 m penetration.

Figure 9.23 graphs resistance and underpressure versus penetration for both High Estimate and Best Estimate k_p, k_f coefficients. The red/thin and blue/thick lines are for reverse punch-through and installation respectively.

Figure 9.23a corresponds to the High Estimate hand calculation, and the left-hand plot (resistances versus penetration) shows the following:

Since $Q_{w,clay}$ is small, installation resistance is almost zero at seafloor. However reverse punch-through resistance is essentially constant, around 10 MN. This is as expected – unit friction values in the adhesion and shear zones are not dissimilar, and the punched-through plug weight remains constant.

At ≈ 2 m penetration, reverse punch-through resistance equals installation. However, this does not imply that reverse punch-through is occurring from this depth onwards – this is due to steel and plug submerged weights assisting and opposing failure respectively.

The corresponding Figure 9.23a right hand plot (underpressures versus penetration) reveals the following:

Underpressures become equal at 3.98 m depth, very close to the base of the clay at 4 m. If H_{clay} had been 3 m, then $\Delta u_{install}$ < Δu_{RPT} at base of clay H_{clay} (i.e. no reverse punch-through failure at all would occur). If H_{clay} = 5 m (was 4 m), then $\Delta u_{install}$ = Δu_{RPT} at 4.65 m depth, again very close to the base of the clay at 5 m.

Underpressure at the base of the clay is just over 100 kPa.

SWP, self-weight penetration, is just under 1.5 m into the 70 kPa clay. This is sufficient to start the suction assisted penetration stage.

Figure 9.23b plots the corresponding results using Best Estimate k_p, k_f coefficients. Since these are lower than the High Estimate values:

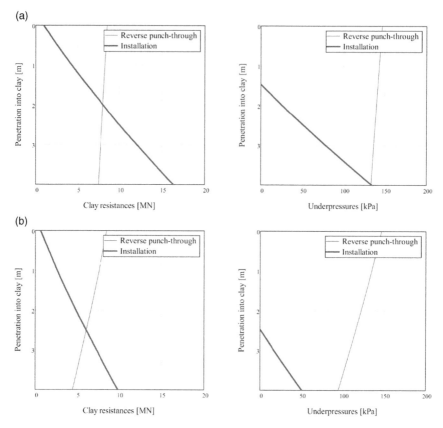

Figure 9.23 Reverse punch-through example. Foundation D = 10 m, WT = 50 mm, W_{sub} = 6 MN. Soil profile: 4 m of ≈ 70 kPa clay over sand. Soil resistances and underpressures versus penetration: (a) High Estimate k_p, k_f = 0.6, 0.05. Reverse punch-through from 3.98 m and SWP = 1.47 m, (b) Low Estimate k_p, k_f = 0.4, 0.03. No reverse punch-through. SWP = 2.48 m.

Both resistances decrease, and SWP increases by 1 m to 2.48 m.

Both $\Delta u_{install}$ and Δu_{RP} drop.

There is no possibility of reverse punch-through. The same conclusion would apply had H_{clay}, the clay layer thickness, been either 3 m or 5 m.

The High Estimate and Best Estimate yf_{RPT} values were –0.01 and –0.39 respectively.

9.11.10.2 Commentary

Based on these results, the probability of reverse punch-through occurring is extremely small. The model excluded water pocket type fluid flow effects in the sand. If they had been included, then increases would occur in the downwards-acting force on the punched-through plug base (currently assumed to be zero) and hence also increase the underpressure Δu_{RPT}. This would make reverse punch-through even more unlikely.

The reverse punch-through example presented here was for competent clay. Had weak NC clay been studied, then (as with using lower k_p, k_f coefficients in the example) the increased SWP values would decrease $\Delta u_{install}$ and the likelihood of reverse punch-through. This exercise is left to the interested reader.

9.12 INSTALLATION IN (WEAK) ROCK

9.12.1 Impact driving

Driving is not the most common way to install a foundation in a full massive of weak rock. Traditionally, weak rock concerns have been encountered mainly for jacket pile or anchor pile installation in cemented carbonate sand (mainly Persian Gulf and Australian projects) or in chalk (CIRIA, 2002). Case studies related to pile driving in other types of weak rock refer mainly to driving through limited thicknesses of few meters (e.g. Puech et al., 1990); see Figure 9.24.

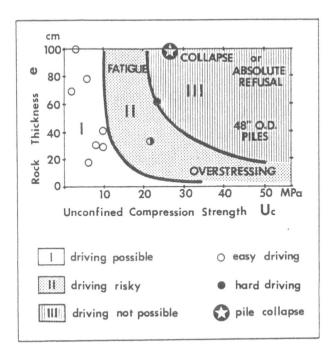

FIGURE 4. DRIVING HAZARDS IN ROCK

Figure 9.24 Impact driving in rock (Puech et al., 1990).

The effect of impact driving on the rock mass will depend on rock characteristics (e.g. strength, porosity, mineralogy and in-situ fractures), but also on the foundation geometry. Stevens et al. (1982) considered that driving in rock will severely fracture the rock layers and reduce the rock to granular material. The shaft friction used in soil resistance to driving predictions (SRD) is therefore defined using sand parameters and end-bearing proportional to the rock unconfined compressive strength (UCS).

In chalk, the above approach proposed by Stevens et al. (1982) proved to be inappropriate (Dührkop et al., 2017 and Wood et al., 2015) for low to medium density chalk. A zone of remoulded material is observed around the pile annulus when driving open-ended piles in these types of chalk. SRD is linked to the grade and density of the chalk and needs to account for "friction fatigue" (unit skin friction decreases with increasing penetration) and set-up effects. Figure 9.25 shows evidence of remoulding and fracturing around a pile after driving in chalk.

In practice, the ground will need to accommodate the volume of steel penetrating. At the start of driving, the easiest way will be to use the available void above the ground, leading to ground uplift around the foundation and

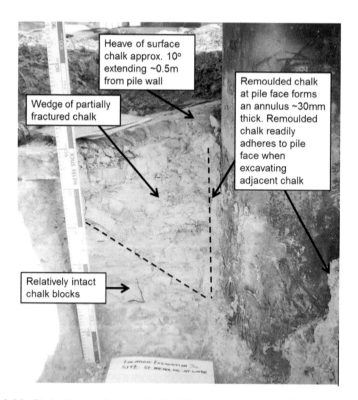

Figure 9.25 Chalk-pile interface model – St. Nicolas pile tests, D = 0.76 m, WT = 44.5 mm (Muir Wood et al., 2015).

creating wedges of fractured material. When the pile starts to penetrate, the ground will accommodate the steel penetration through a combination of:

- Rock crushing around the pile wall. Depending on the rock type, the crushed material will be either in a pasty or remoulded state (e.g. for grade C and D chalk) or as a granular material.
- Radial fractures up to a few foundation diameters from the foundation edge. In the presence of natural fractures, blocks will be pushed radially and fractures will tend to close.

Parameters such as rock porosity, geological strength index (GSI) or presence of underlying weaker layers are not considered in usual empirical methods to derive soil resistance to driving. These parameters will, however, play a fundamental role in the way the ground will accommodate the foundation penetration.

Driving installation of intermediate foundations in weak rock will not differ much from classical flexible pile installation. It should however be noted that for large diameter foundations, steel integrity will need to be ensured by either the use of thick plates (driving shoe) or internal stiffeners (see Figure 9.26). Depending on the foundation diameter and risk of buckling (see Section 9.13.5), the bottom plate thickness can be in the range of 70 to 130 mm.

9.12.1.1 Pile driving refusal

There are no clear criteria to define drivable or undrivable conditions. The experience shows that calcareous weak rock or cemented sand can be driven as long as a cone penetrometer can penetrate. High-density chalk will tend to behave as limestone and would not necessarily be easy to drive through.

API RP 2A (API, 2000) provides some guidance on the maximum blow count not to exceed to ensure pile integrity. Other contractor specifications go a bit further and include lower criteria in terms of maximum percentage of yield stress to not overcome in rock. Criteria limiting the reflection ratio measured by pile driving monitoring are also used.

It should be noted that chalk can experience significant increase of shaft resistance with time (i.e. set-up). Driving interruption of few hours can increase the soil resistance to driving and therefore create a refusal condition by increasing the blow count and reflection ratio.

9.12.1.2 Risk of buckling

Open-ended piles are vulnerable to tip distortion. An initial imperfection from the theoretical pure cylindrical geometry can progressively grow with increasing pile penetration when the stiffness of the surrounding soil exceeds the elastic pile stiffness (Figure 9.27). Care is required during fabrication, transportation and installation to avoid creation of an initial imperfection

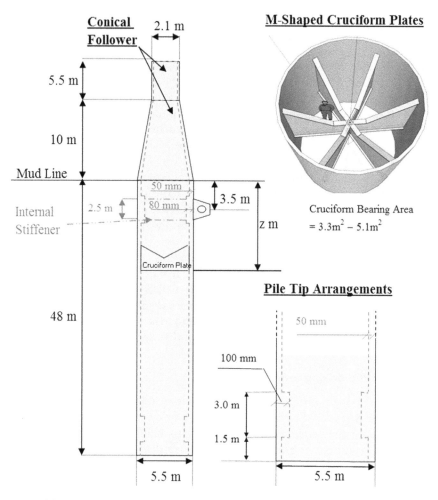

Figure 9.26 Ichthys pile tip stiffening arrangement, D/WT ratio over 110 (Erbrich et al., 2017).

that can lead to buckling. Aldridge et al. (2005) present equations, and both Erbrich et al. (2010) and Aldridge et al. (2005) describe conditions for propagation of an initial imperfection during pile penetration (i.e. extrusion buckling). The propagation of a small initial deformation requires that the soil be both strong enough and stiff enough to deform the pile further.

For monopile or large diameter anchor piles, the ratio between the diameter and the pile tip thickness (i.e. D/WT ratio) is becoming very high compared to classical flexible piles (see Figure 9.28). Figure 9.13 shows that for D/t greater than 40, buckling was observed. Risk of insufficient hoop stiffness or heterogeneous conditions across the diameter are significantly greater for stubby large diameter foundation. The ring stiffness is inversely proportional to cube of D/t ratio.

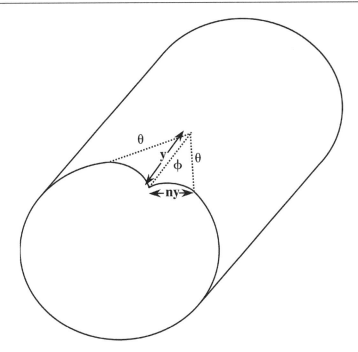

Figure 9.27 Plastic hinge mechanism assumed to cause local tip buckling (Aldridge et al., 2005).

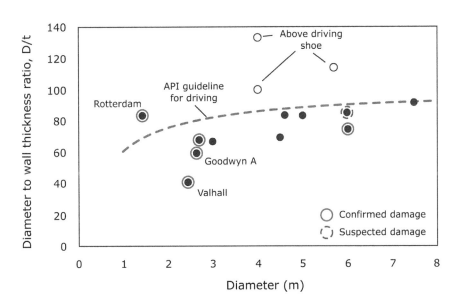

Figure 9.28 Pile tip geometries, oil and gas full blue dots and monopiles empty dots, red circles correspond to recorded tip damage (Randolph, 2018).

Particular care needs to be accounted for in the case of heterogenous conditions around the pile tip (e.g. inclined hard layer, boulders or flint beds). Presence of strong nodules in a softer matrix can initiate localised damage if not easily pushed aside or split (Holeyman et al., 2015). The buckling risk can be partially mitigated by considering a driving shoe (i.e. increase pile tip thickness up to 130 mm) end adapted driving procedure.

Classical monitoring with strain gauges and accelerometer placed at the top of the pile are not sufficient to issue real-time warning at the start of buckling initialisation.

A careful reading of monitored information could indicate, post-installation, where the buckling started. A modification of the impedance from the starting damaged point should be required to match the measured signal. Proper real-time monitoring would require sensors to be placed at the toe of the pile, but the installation of sensors would risk weakening the pile (e.g. creation of a groove to install fibre-optic strain gauges). The use of fibre optics without a groove (i.e. only fixed with glue) is under development and seems to be a promising approach.

Special procedures (i.e. decision tree) can be applied in homogenous or well-known stratigraphy to interpret in real time the soil resistance to driving (SRD) evolution and take appropriate mitigation measures (e.g. drilling inside the foundation down to the target penetration depth). However, in case of heterogeneous material, it will be difficult to differentiate if the observed increase in blow count (i.e. increase of tip resistance) is due to a modification of the rock strength or an increase of the tip area induced by deformation. In practice, the occurrence of buckling is often discovered only when drilling inside the pile is required.

9.12.2 Drilling

9.12.2.1 Drive drill drive

Drilling can be used as a contingency measure when driving is expected to be difficult (as the drilling tool is generally already mobilised together with the hammer) or in case of unexpected pile refusal (e.g. presence of boulder). After the end of driving, the drilling head is inserted inside the pile and drilling starts from the seafloor. Drilling can either stop after the particularly difficult layer (or boulder) or be continued to the pile target depth. In the latter case, hole stability needs to be guaranteed. The driving process is then continued until reaching the target depth or the next difficult layer.

Drive Drill Drive (DDD) operations often require under-reaming capability, i.e. the possibility for the drilling head to enlarge its diameter once it reaches the pile tip. Drilling more than the pile diameter removes resistance beneath the tip in a particularly hard layer (or boulder).

The switch of tool between the hammer and drilling tool is time consuming. Drive Drill Drive (DDD) installation is often scheduled only if one pass of drilling is expected to be sufficient. If several passes are expected, a drilled and grouted process might become more cost effective.

9.12.2.2 Drilled and grouted

When driving is not an option or when the risk of pile buckling is too high, an alternative can be to drill and grout the foundation. The drilled hole diameter is equal to the foundation diameter plus two times the grout annulus. The foundation is then lowered in the hole and the void between the foundation and the ground is grouted from the bottom to the top.

Open drilling can start from the seabed when the rock mass is stable enough. In most cases, upper sediments or degraded rock are present at the seabed, and drilling needs to be advanced using temporary (i.e. retractable) or sacrificial (i.e. stay in place after drilling operation) casing supporting unstable ground until the grouting operation.

Figure 9.29 shows an example of a drilled and grouted installation sequence using a casing to equalise the overburden layer and support the top drilling equipment.

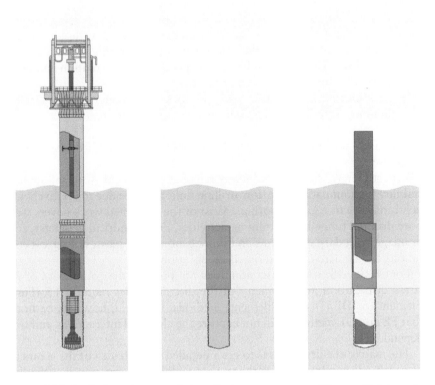

Figure 9.29 Drilled and grouted installation sequence with sacrificial casing (Courtesy of Fugro).

The main risks during drilling are:

- Hole collapse: meaning retrogressive failure of the wall resulting in damage to, or loss of drilling equipment and requiring relocation of the foundation;
- Over-break: meaning local instability of the hole wall, likely resulting in an increase in hole diameter and potentially a deterioration of the rock condition local to the hole wall;
- Decrease of hole diameter: preventing the foundation being lowered and any temporary casing being removed (e.g. swelling rock);
- Hole inclination out of tolerances;
- Drilling rate much lower than expected (e.g. in presence of large boulder or plastic clay if drilling tool is not designed for the conditions encountered).

Before drilling, the rock mass is in a state of equilibrium. After drilling, instability can occur if:

- the post drilling redistributed stress-state exceeds rock strength.
- a network of closely spaced discontinuities could lead to the sliding of blocks. The presence of at least three discontinuities passing through the drilled hole is necessary for a block to slide (two discontinuities could be sufficient, but the risk of getting an unstable block would be quite low). The risk of sliding will depend on the orientation and properties of joints and the rock properties.
- rock-chemical interaction leads the rock to weaken or swell. Drilling fluid needs to be carefully selected, particularly in shale.

In all cases, the time during which the hole is kept unfilled needs to be minimised.

Unfortunately, there are not many references applicable for shallow hole stability for foundations (i.e. low drilling fluid pressure due to relatively low depth relative to e.g. well-boring). Most of the published work covers well-bore drilling, tunnelling or mining activities. For cylindrical geometry, zero drainage boundary conditions and homogenous rock mass, hole stress conditions can be derived based on elastic solutions.

For weak rock where discontinuities are present, the complexity of the stress distribution and geometry of the discontinuities requires FE (Finite Element) or DE (Discrete Element) methods. It should, however, be noted that FE and DE methods will not capture the chemical interactions and time dependency.

The main difficulty remains to get a detailed enough idea of the geometry of nearby discontinuities. Offshore, the available information is often limited to the logging of joints and fractures on 100 mm samples from a single borehole. It might be necessary, where faults are prevalent, to perform several boreholes per foundation, enlarge the depth of boreholes (to compare geological markers) and/or consider downhole logging such as natural gamma-ray, televiewer and calliper. These tools can provide valuable information on the nature, frequency and inclination of discontinuities (crossing the borehole).

During grouting operations, it will be necessary to avoid a grout flow inside the foundation. This can be done by either creating a grout plug at the bottom of the foundation before filling the annulus, through a physical boundary (e.g. internal membrane) or through a mechanical closure (e.g. using external packers).

In the presence of fractured rock or karstic voids, the volume of grout to be injected can largely exceed the annulus volume (volume between the foundation and the rock). Grouting operations offshore remain a delicate operation that is difficult to control.

9.12.3 Vibratory

Pile installation using vibratory means would not be considered if the stratigraphy includes cemented material or hard (even small) layers.

9.13 PRESENTATION OF INSTALLATION ASSESSMENT

Presentation of suction foundation installation analyses usually includes:

- soil layering and parameter values
- foundation geometry
- preload values
- Best Estimate and High Estimate resistances versus depth
- Best Estimate and High Estimate underpresssures versus depth, including allowable/limiting pressures for base failure and liquefaction.

Figure 9.10 has given an example for an anchor pile in LOC clay. Figure 9.17 presents a typical example for caisson installation in sand. Note the following:

- Double plot (resistance R and underpressure derivative Δu versus depth z): the latter plot cannot be used if the preload alters. Entering the $R - z$ profile with W_{steel} value gives self-weight penetration. Entering the $\Delta u - z$ plot with the target penetration depth gives the maximum underpressure (usually).
- For clays, the $\Delta u - z$ plot also shows Δu_{all}, the allowable suction pressure to prevent base failure.
- For sands, the $\Delta u - z$ plot also shows Δu_{liq}, the underpressure causing sand plug liquefaction.

- Other allowable pressures may be necessary, for example water depth and structural criteria
- High Estimate results have been omitted for clarity – normally both Best Estimate and High Estimate results are provided.

9.14 RETRIEVAL AND REMOVAL RESISTANCE ASSESSMENTS

9.14.1 Suction foundations

Procedures for retrieval and removal are similar to those for installation (Section 9.2 on penetration resistance assessment), except that:

- set-up may have occurred in normally consolidated clay (Section 9.10.3) and also in sand (Section 9.11.3)
- overpressure (downwards seepage in soil plug) increases inner skin friction in sand
- end-bearing resistance at tip level is neglected, provided that there is no passive suction below the top plate and the vent is open
- resistance (both friction and end-bearing) may occur *above* internal friction breakers
- dead weight counteracts overpressure
- special techniques (e.g. Broughton et al., 2002) are needed to apply overpressure beneath underbase grouted foundations.

9.14.2 OWT monopiles

Driven monopiles cannot be removed in their entirety – their pull-out capacity (resistance plus submerged weight) is significant. For example, a 5 m diameter 30 m embedded length monopile pull-out capacity could easily exceed 60 MN, depending on the exact soil profile and other factors. There are, however, on-going developments to investigate removal of such piles, see e.g. Balder et al. (2020). Marine legislation generally stipulates that piles be cut off at a certain depth (typically between 3 m and 5 m) below seafloor. This is generally achieved by explosives, mechanical or abrasive cutters. Jetting may be necessary to remove the soil beforehand. Pile cutting may be either internal or external; usually the former approach is selected, in view of the soil volumes to be excavated. Grout overflow or other seafloor obstructions may hinder access.

Drilled-and-grouted monopiles cannot be removed entirely either. For the decommissioning phase, the requirement is generally to cut the monopile 1 m below seabed and to fill it with rock. Only a few offshore windparks have already been decommissioned. Few relatively small drilled and grouted monopiles (around 3.5 m diameter) were successfully cut 1 m below seabed using an internal jetting system (e.g. Blyth monopiles).

Vibratory driven monopiles can also be cut off using the same driven pile methods. Theoretically, vibratory methods could also be used, using a higher energy vibratory hammer due to set-up in sand. However, at the time of writing, this method has not been applied offshore.

9.15 CLOSURE

This lengthy chapter has covered intermediate foundation installation. Most sections are devoted to suction-assisted foundations. This is because they are more challenging and complex than monopile foundations, for which design methods are well understood and accepted by industry.

At the time of writing there is no general agreement/consensus as to which design methods/models are appropriate for suction foundations. In terms of increasing maturity, installation and penetration resistance methods are more mature for clays (especially NC clay), than sands, followed by mixed sand/clay profiles. Reasonably accurate models are given for these profiles.

However, installation in clay is less problematic than in sand – the sand model accuracy needs to be improved (e.g. Figure 9.17) and field data suggest that i_{crit} may be exceeded (e.g. Figure 9.15). In addition, little is known about "set-up" factors for skin friction in sand. Hence, more research, calibrated using high-quality field data, is required on installation in sand, where plug liquefaction and plug heave are critical issues. At the moment, not exceeding the limiting $i_{crit} = \gamma_{sub}/\gamma_w$ value gives L/D values between 1.0 and 1.5, the latter usually associated with low relative density layers near seafloor. Geotechnical engineers prefer to embed foundations as deep as possible: soils get more competent with depth, and cost savings accrue. Hopefully, the observational method, together with the research described here, will eventually lead to achieving higher intermediate foundation embedments L/D in sand.

Pseudo code for three simple programs (and verification input/output data) have been provided: *install_clay, install_sand* and *water_pocket* for clay over sand. These simple building blocks can be modified and combined for solving real-world suction assisted foundation installation problems.

Chapter 10

In-place resistance

10.1 INTRODUCTION

This chapter covers in-place resistance for both support type and anchor pile type intermediate foundations.

10.2 LOADING CONDITIONS AND SOIL RESPONSE

In routine design, T (torsion) loads are small compared to significant HM (lateral) loading. In addition, offshore extreme loading conditions are such that soils – even sands – can usually be modelled with an undrained response (Chapter 5).

10.3 IN-PLACE FAILURE MODES

Figure 2.2 (Section 2.2, on in-place resistance modes) showed six failure modes for intermediate foundations under combined VHM loads in *undrained* soil. All assume co-planar HM (horizontal load and moment) and negligible T (torsion). Figure 2.3 showed the corresponding failure modes for *drained* soil response. All modes are suitable for preliminary/routine design calculations, but possibly need to be complemented by FEA for final design.

Figures 2.2 and 2.3 are both for *combined* VHM loading. Figure 10.1 (Section 10.5 on maximum axial resistance) gives the corresponding failure models for *pure axial (pull-out)* V load.

Resistance envelope models for VHM(T) loading are discussed further in Section 10.10.

10.4 TENSION CRACKS AND GAPPING

Tension cracks (also known as gaps) can occur behind the trailing edge (active side) of foundations under lateral load. This has been realised as a potential problem for sheet pile walls for many years. The crack depth depends upon many factors, including the lateral load magnitude and the in-situ stress profile.

The occurrence of either tension cracking or gapping reduces VHMT resistance. For offshore foundations, if a tension crack develops to skirt tip level, then in-place undrained resistance capacity at tip level (including reverse end-bearing) may be severely affected by loss of suction in fine-grained soil. In addition, in the zone above the tension crack, a small reduction in bulkhead net resistance (passive minus active) occurs.

Tension cracking is not generally considered for normally consolidated clay subjected to short-term loads (Randolph and Gourvenec, 2011). Jeanjean et al. (2006) found no evidence of gapping in more competent OC clay ($s_u \approx 25$ kPa at 9 m depth). Gapping is, however, expected to be non-reversible in cemented formations or rock. In the case of drilled and grouted piles, the gap will most likely occur between the steel pile and the grout annulus. This gap should normally be limited to the upper 1–2 meters due to the limited displacements.

For intermediate anchor foundations in normally consolidated clay, tension cracking is not usually considered. This is because locating the lug slightly below optimum depth causes restoring loads and decreases the risk of forward rotation. However, tension cracking and gapping may need consideration for intermediate support foundations under overturning loads and/or in stiff clays.

Mitigation measures for foundations include a flexible mat (a "gap arrester" or "mud liner") around the perimeter (e.g. Keaveny et al., 1994; Mana et al., 2013). This inhibits a tension crack by preventing water supply, and hence suction loss, at the soil–foundation interface.

10.5 MAXIMUM AXIAL RESISTANCE - SUCTION PILES

10.5.1 Failure modes for maximum axial tensile resistance

This section discusses possible axial failure modes, and selection of the most appropriate mode. The following sections discuss appropriate unit friction and unit end-bearing parameter values for undrained "clay", undrained "sand" and drained "sand" respectively. Skin friction and end-bearing contributions to axial resistance are assumed to be uncoupled.

Distinctions should be made between:

V loads (tensile or compressive, covered in Section 3.2 on intermediate foundation types) soil response (undrained or drained, covered in Section 5.2 on drained-undrained-partially drained).

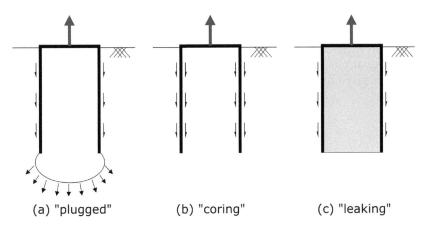

Figure 10.1 Failure models for maximum axial tensile resistance, V_{max}, in undrained soil. Reverse end-bearing is included only in the "plugged"
model (Senders and Kay, 2002).

Figure 10.1 shows the three possible failure modes – "plugged", "coring" and "leaking" – under tensile V loads. Identification of these modes have been based on previous work. Fuglsang and Steensen-Bach (1991) reported results of 1 g model and 40 g centrifuge pull-out tests on 65 mm and 80 mm diameter L/D = 2 caissons in kaolin clay. To derive end-bearing and inner/ outer skin friction components, they used three model caisson types: "plugged" ($F_o + Q_{base}$), "coring" ($F_o + F_i$) and "coring without the clay plug" (F_o). They verified that there was indeed reverse end-bearing failure. In several 1 g "plugged" tests, but none of the centrifuge tests, brittle tensile failure was observed – the clay at the caisson tip abruptly failed in tension. This was possibly the "leaking" mode – the tests were carried out underwater.

Randolph and House (2002) considered the same three "plugged", "coring" and "leaking" failure modes as Fuglsang and Steensen-Bach (1991), using instead the term "base-vented" when a hydraulic short circuit (i.e. "leaking") prevents underpressure at the caisson base. For the "coring" mode, reverse end-bearing resistance (Q_{wall}) under the caisson wall is included. Although admittedly small, it is optimistic to include this component. Senders and Kay (2002) presented the same three models, but with component Q_{wall} omitted from the "coring" mode. The following paragraphs give fuller details and update the aforementioned work.

10.5.1.1 Tensile V loads (anchor foundations) in "undrained" soil

If the top plate vent is permanently sealed, then reverse end-bearing ("passive suction") can be relied upon for a limited time period, such as peak pull-out loads during a storm. In this case, axial resistance (V_{max}) equals outer friction and reverse end-bearing. This is the "plugged" failure mode. Note that the tensile resistance at foundation tip level is given by classical bearing capacity theory as "$A_o s_u N_c - A_o q$", q being the surcharge. However, the term "$A_o q$" is equal and opposite to the soil plug self-weight, W_{plug}. Hence, a W_{plug} term does not appear in the expression for V_{max}, the axial resistance at pile head level. Hence, "plugged" V_{max} and V_{load} (axial pull-out load) are given by the equations:

$$V_{max} = F_o + Q_{base} = F_o + N_{c,circle} s_{u,av,L} A_{base} \tag{10.1}$$

$$V_{load} = F_o + Q_{base} + W_{steel} \tag{10.2}$$

If reverse end-bearing ("passive suction") cannot be relied upon, then either the "coring" or "leaking" failure mode can occur. Both modes imply some form of hydraulic leak and zero pressure under the top plate. Examples of design situations that readily spring to mind include:

(a) A top plate seal cannot be guaranteed to function perfectly during (say) 30 years sustained tension load, and, due to leakage, the pressure under the top plate will eventually reduce to (almost) zero.
(b) A hydraulic leak occurs at (or just below) foundation tip level due to a seam or layer of permeable material in the vicinity of the foundation tip which behaves essentially hydrostatically (no excess water pressure) and provides zero reverse end-bearing. Note that seams, between 20 mm and 76 mm thick according to ASTM 2487 (ASTM 2011) and ASTM 2488 (ASTM 2009), may be undetected by CPTu. Moreover, in the horizontal direction, a 20 mm thick fine sand seam (permeability $k = 10^{-4}$ m/s) has 40 times the transmissivity of a 5 m thick clay layer ($k = 10^{-8}$ m/s).
(c) A hydraulic leak occurs at foundation tip level due to a gap extending from seafloor to tip level caused by excessive foundation rotation of an anchor pile with a wrongly located pad-eye. Again, this results in zero end-bearing resistance.
(d) Another possible scenario is due to anchor chain trenching, which may adversely influence reverse end-bearing resistance, usually relied upon for FPSO mooring anchors. To help ensure backwards rotation, design typically includes a deeper anchor lug level than is 'optimal', and this further reduces the distance between lug and tip. If the chain trench extends to lug level, then there is an increased risk of either (i) a reduced drainage path length H (H = tip to trench base, was H = tip to seafloor), or (ii) a hydraulic leak occurring between lug and tip (i.e. seawater breaking tip suction). Both situations could result in hydrostatic water pressure at foundation tip level during the design lifetime). Hence, as a consequence of trenching, reverse end-bearing was cautiously omitted by Alderlieste et al. (2016).

In the "coring" failure model, there is no reverse end-bearing under the tip wall, and axial resistance (V_{max}) equals outer friction (F_o) and inner friction (F_i). The corresponding V_{load} (axial pull-out load) equals $F_o + F_i + W_{steel}$. In the "leaking" failure model, V_{max} simply equals outer friction (F_o), and V_{load} (axial pull-out load) equals $F_o + W_{steel} + W_{plug}$.

Hence, "coring" V_{max} and V_{load} are given by the equations:

$$V_{max} = F_o + F_i \qquad (10.3)$$

$$V_{load} = F_o + F_i + W_{steel} \qquad (10.4)$$

and "leaking" or "plugged" by the equations:

$$V_{max} = F_o \qquad (10.5)$$

$$V_{load} = F_o + W_{steel} + W_{plug} \qquad (10.6)$$

Design axial tensile resistance is usually taken to be one of the following:

if reverse end-bearing ("passive suction") can be relied upon:
"plugged" V_{max} resistance
    if reverse end-bearing ("passive suction") cannot be relied upon:
if there is no possibility of the "leaking" failure model occurring:
    "coring" V_{max} resistance
if the "leaking" failure model can occur:
    minimum of the "coring" and "leaking" V_{max} resistances.

Note that for a normally consolidated clay ($s_u/\sigma'_{vo} \approx 0.25$) with clay-steel inner adhesion factor $\alpha_i \approx 0.6$, "leaking" is more critical than "coring" failure when embedment ratio $L/D \approx 3$. That is, if "leaking" failure can occur, then "coring" resistance is probably governing for short, stubby foundations, whereas "leaking" is more likely to be critical for longer foundations.

Finally, for routine FEA, it is very difficult to model pull-out using either a "plugged" failure model with a user-assigned N_c value (i.e. reduced REB; Clukey and Morrison, 1993; Clukey et al., 1995), or a "leaking" failure model. In the remaining "plugged" and "coring" cases, design practice should be to obtain and interpret the sum of soil forces along the inner and outer shaft areas (i.e. mobilised friction F_o and F_i) and over the base area (i.e. mobilised bearing resistance Q_{base}). If "plugged" failure is occurring, then one would expect $V_{load} \approx F_o + Q_{base}$ and $V_{load} \approx F_o + F_i$ if "coring". For the "plugged" condition, the back-figured $N_{c,circle}$ value should be compared with the Skempton (1951) value for circular foundations. In particular, $N_{c,circle}$ is expected to be in excess of 9 for $L/D = 2.5$, for the reasons given in Section 10.5.2.

10.5.1.2 Compressive V loads (support foundations) in "undrained" soil

For compressive V loads, the "leaking" failure mode does not occur and the number of failure modes reduces from three to two (i.e. "plugged" and "coring"). W_{steel} acts in the opposite sense: it reduces the design compressive load. In addition, V_{wall} (wall tip end-bearing resistance) is added to the "coring" failure mode. It is seen that these aspects are similar to compressive offshore pipe pile design procedure.

However, design axial compressive resistance depends on whether or not there is "top plate bearing" (i.e. soil plug or grout is in contact with the top plate underside):

if "top plate bearing" can be relied upon:

"plugged" V_{max} resistance

if "top plate bearing" cannot be relied upon:

minimum of the "plugged" and "coring" V_{max} resistances.

This procedure is unlike conventional pipe pile design. The difference in procedure derives from from the differing soil plug lengths L_{plug} (both in clay and sand) for pipe piles and intermediate foundations. Pipe pile L_{plug} values are usually somewhat less than L (pile embedded length); this is mainly due to inertial effects during installation by impact driving (Smith and Chow, 1982; Chow, 1981). Typically, for both sands and clays, $L_{plug} \approx 0.9$ L. Hence, there is generally no "top plate bearing" for pipe piles. On the other hand, intermediate foundations installed by suction in clay causes soil plug heave to match part of the soil volume displaced by the foundation steel. Section 9.10.2 (on plug heave in clay) suggests that for clays, $L_{plug} \approx 1.02$ L. Hence, for suction-installed intermediate foundations in clay, there is generally "top plate bearing", and the "plugged" failure model is appropriate.

Table 10.1 summarises the V_{max} and V_{load} equations.

Table 10.1 Maximum Axial Resistance and Load in Undrained Soil.

Axial load	Failure model	Maximum axial resistance V_{max}	Maximum axial load V_{load}
Tension (anchor foundation)	"plugged"	$F_o + N_c\, s_{u,av,tip}\, A_{base}$	$F_o + N_c\, s_{u,av,tip}\, A_{base} + W_{steel}$
	"coring"	$F_o + F_i$	$F_o + F_i + W_{steel}$
	"leaking"	F_i	$F_i + W_{steel} + W_{plug}$
Compression (support foundation)	"plugged"	$F_o + N_c\, s_{u,av,tip}\, A_{base}$	$F_o + N_c\, s_{u,av,tip}\, A_{base} - W_{steel}$
	"coring"	$F_o + F_i + N_c\, s_{u,tip}\, A_{wall}$	$F_o + F_i + N_c\, s_{u,tip}\, A_{wall} - W_{steel}$
	"leaking"	Not applicable	Not applicable

Note(s): See Figure 10.1 for tensile failure models

10.5.1.3 Example – maximum axial resistance

To illustrate the equations and to examine the sensitivity to the three failure modes, embedded anchor lengths L have been obtained for a constant diameter D = 5 m and 7500 kN factored tensile pull-out load. Results in Table 10.2 used the following data:

$s_u = 2 + 1.5 z$ [kPa, m], $\gamma_{sub,soil} = 6$ kN/m^3

$D = 5$ m, $D/WT_{side} = 200$, $D/WT_{top} = 100$, $L_{plug} = L$, $\gamma_{sub,steel} = 67$ kN/m^3

$\alpha_i = \alpha_o = 0.65$

$\alpha_{D,su} = 0.5$

$\lambda_{L,Bw} = 1.0$.

Table 10.2 shows that (for the data presented here):

- there is a progressive L increase – embedment ratios L/D are ≈ 3, 4 and 4.5 for the "plugged", "coring" and "leaking" failure models respectively
- "plugged" is the most efficient – soil and drainage conditions permitting $B_{w,caisson}$ for "plugged" is 0.75 of the corresponding "leaking" value
- all L/D ratios are < 5, and hence amenable to suction installation.

Table 10.2 Foundation Embedded Length L for 5 m Diameter Anchor, 7500 kN Factored Tensile V_{load} in Undrained Soil and Varying Failure Modes.

Parameter		Failure Mode		
		"plugged"	"coring"	"leaking"
L	[m]	15.64	20.01	22.33
L/D	[–]	3.13	4.00	4.47
Resistances				
F_o	[kN]	2192	3474	4274
F_i	[kN]	Not applicable	3439	Not applicable
Q_{base}	[kN]	4830	Not applicable	Not applicable
V_{max}	[kN]	7022	6914	4274
Loads				
$B_{w,caisson}$	[kN]	475	589	651
$B_{w,plug}$	[kN]	Not applicable	Not applicable	2578
$B_w = B_{w,caisson}$ $+ B_{w,plug}$	[kN]	475	589	3229
resistances + loads				
$V_{load} = V_{max}$ $+ B_w$	[kN]	7500	7504	7503

Note(s): H = M = 0.

Note that these L values are solely for pure V load. Placing the lug level at optimum depth below seafloor for resisting combined load gives significantly smaller L values – between 12 m and 14 m (see the Example in Section 10.7).

10.5.2 Undrained ("clay") soil response

Skin friction resistance and Pile-soil adhesion coefficient α:

- α values are generally lower than recommended by ISO 19901-4 (ISO 2016a), Section 8.1.3 (Skin friction and end-bearing in clay soils) for driven pipe piles. In NC clay, α is around 15–20% lower.
- α self-weight penetration zone may be higher than in the suction installation zone.
- Outer skin friction may be higher than and inner skin friction (e.g. Jeanjean, 2006; Jeanjean et al., 2006).
- Cautious values may be $\alpha_o = \alpha_i = 0.65$.
- Reduced α_i (e.g. $\alpha_i = 0.3$) may be appropriate above internal stiffeners in clay due to remoulding ($\alpha_i = 0$ if extruded above stiffener).

End-bearing resistance and Bearing capacity factor N_c:
  For constant shear strength with depth, the equations are:

$$N_{c,strip} \approx 7.5 \text{ for wall end bearing (strip) (Skempton, 1951)} \qquad (10.7)$$

$$N_{c,circle} \approx \min\left(6\left(1+0.2\left(L/D\right)\right), 9\right) \text{ (Skempton, 1951))} \qquad (10.8)$$

$$N_{c,circle} = \min\left(6.2\left(1+0.34\arctan\left(L/D\right)\right), 9\right) \text{ (DNV, 2018)} \qquad (10.9)$$

Linearly increasing shear strength with depth:

- For (reverse) end-bearing over the full foundation base area, a reasonable approach is to take $N_c = N_{c,circle}$ and s_u averaged over (say) 0.25 D below tip level. This approach is consistent with that used for jack-up spudcan leg penetration assessments, see. ISO 19905-1:2016 (ISO 2016b), Section A.9.3.2.2. Other organisations may have different opinions regarding averaging depth etc.
- For (reverse) end-bearing over the foundation wall tip area, take $N_c = N_{c,strip}$ and s_u at tip level.
- Other methods are available for *shallow* foundations; see Section 3.2.1 on in-place resistance modes.

Note that it is cautious to assume that bearing capacity factor $N_{c,circle}$ has a maximum value of 9 for closed-ended and plugged open-ended piles. See Section 10.5.5 on axial myths.

10.5.3 Undrained ("sand") soil response

Assessing tensile resistance of undrained sand under transient load is extremely difficult. Considerations include:

- A key aspect is suction development under top plate.
  - equivalent s_u is very sensitive to sand dilatancy parameter D (e.g. DNV, 1992)
  - suction is limited by pore fluid cavitation (e.g. McManus and Davis, 1997)
- Additional considerations include:
  - pore water pressure build-up and dissipation
  - reliability of soil parameters (e.g. k, c_v and S-N curve)
  - stress redistribution (especially outer skin friction) from skirt to plug base
  - sand plug loosening due to installation
  - sand plug re-compaction (if any) during previous environmental (storm) loading
  - amount of compressive preload applied prior to (single) tensile wave load
  - V resistance reduction due to HM loads (coupling).

Some examples of evidence for tensile resistance in sand include:

- *Field*: Draupner jacket, New Year's storm, 1 January 1995, piezometer measurements indicated that the bucket foundation could resist short duration loads without any problems (Svanø et al., 1997, Hansteen et al., 2003).
- *Laboratory*: Pressure chamber (and onshore field) testing of suction caissons under tensile loading in sand (Kelly et al., 2003, 2004, 2006a, 2006b, Houlsby et al., 2005).
- *Numerical*: Transient cyclic loading of suction caissons in sand (Cerfontaine et al., 2016).

Commentary:

- These challenges tend to disappear with increasing embedment ratio (think pipe piles) BUT suction installation when L/D > 1 is difficult.
- This is very different to the clay–sand separation challenge described in Section 9.12.2 on the water pocket model.

- If concerns exist about high tensile axial resistance, then permanent passive suction (Section 12.5) could be considered.

10.5.4 Drained ("sand") soil response

10.5.4.1 Compressive loads

If the failure mode is "plugged", then V_{max} is given by:

$$V_{max} = \gamma_{sub}L^2 / 2\beta_o \pi D + \gamma_{sub}L\,N_q A_{base} \qquad (10.10)$$

And, if it is a "coring" failure mode, then we have:

$$V_{max} = \gamma_{sub}L^2 / 2(\beta_o + \beta_i)\pi D + \gamma_{sub}L\,N_q A_{wall} \qquad (10.11)$$

Table 9.6 gives values of ISO 19902 pile bearing capacity factor N_q.

10.5.4.2 Tensile loads

Assuming a "coring" failure mode, and ignoring reverse end-bearing below the wall tip, then V_{max} is given by:

$$V_{max} = \gamma_{sub}L^2 / 2(\beta_o + \beta_i)\pi D \qquad (10.12)$$

Note that friction resistance for intermediate foundations using ISO 19901-4 (2016) for driven piles is considered to be optimistic. The reasons for decreased friction include:

(a) less densification (lower energy and soil volume disturbance)
(b) inner friction reduced due to sand plug loosening (suction installation)
(c) outer friction reduced due to installation (vibratory installation only).

10.5.5 Axial myths

10.5.5.1 Bearing capacity factor N_c

It is a geotechnical myth that $N_{c,circle}$ has a maximum value of 9 (Skempton, 1951) for closed-ended and plugged open-ended piles. This value is cautious. The DNV (2005) equation is merely a "better" (non-linear) fit to the original Skempton (1951) Figure 2 curve. Skempton's work was based on theoretical considerations, small-scale laboratory, and two field tests on screw cylinders/screw piles. All these boundary conditions are not

Table 10.3 Bearing Capacity Factor $N_{c,\,circle}$ – Numerical Analyses.

Boundary Condition Feature	$N_{c,\,circle}$ Reference				Remarks
	Skempton (1951) Figure 2	Griffiths (1982) Figure 4	Salgado et al. (2004) Figure 11	Edwards et al. (2005) Figure 3	
Open/closed hole	Open	Closed	Closed	Closed	Griffiths (1982) Figure 5 (open hole) ≈ same as Skempton (1951) curve
Smooth/rough sides	Not applicable (field/ laboratory/ theoretical)	Smooth	?	Smooth	
Load/ displacement control base		Displ	Pressure	Displ	
Rough/smooth base		Smooth	Rough	Rough	

necessarily the same as for (plugged) pile/intermediate foundation tip end-bearing. Table 10.3 lists more recent numerical analyses using both Limit Equilibrium and Finite Element Analysis. For a given tip penetration, the $N_{c,circle}$ value is influenced (increases) with increasing pile-soil adhesion coefficient α_o, tip roughness and plug rigidity. Figure 10.2 compares the $N_{c,circle}$ – L/D curves for these various conditions.

For comparison, 2D axisymmetric FEA of CPT penetration, a 60° cone, gives $N_k \approx 12.5$ (van den Berg, 1994). In addition, the more recent Figure 10.2 $N_{c,circle}$ data are valid only for "closed-hole" situations (i.e. closed-ended and plugged open-ended). Lower values should be used for "open-hole" situations. Incidentally, this explains the lower $N_{c,spudcan}$ values (Houlsby and Martin, 2003 and ISO 19905-1:2016, Annex E1): an "open hole" situation exists above the spudcan to permit backflow.

Physical confirmation for $N_{c,circle}$ exceeding 9 has also been obtained from a number of centrifuge tests on "plugged" caissons. Fuglsang and Steensen-Bach (1991) performed both 1 g laboratory tests and 40 g centrifuge tests on L/D = 2 caissons to verify the presence of reverse end-bearing failure. They obtained average $N_{c,circle}$ values of 9.2 (weak 10 kPa clay) and 8.1 (stronger 20 kPa clay), i.e. both in good agreement with the Skempton (1951) value of 8.5 for circular foundations. Clukey and Morrison (1993)

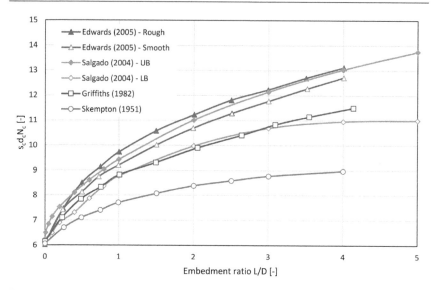

Figure 10.2 $N_{c,circle}$ versus embedment ratio showing dependency on various boundary conditions (see Table 10.3).

reported an $N_{c,circle}$ factor of approximately 11 for caissons with L/D = 2 in NC clay. This factor may be optimistic (see e.g. Randolph and House, 2002). Randolph and House (2002) found $N_{c,circle}$ to be between 14 and 15 for short term pull-out load applied to L/D = 4 caissons in NC clay. Chen and Randolph (2007b) studied caissons with prototype D = 4 m, L = 12 m (i.e. an embedment ratio L/D = 3) in kaolin clay. For NC clay, inferred combinations of average α_o and $N_{c,circle}$ values were (0.90, 11.0) or (0.76, 12), with α_o > 1 for $N_{c,circle}$ of 10 or less. For LOC clay, the corresponding average α_o and $N_{c,circle}$ combinations were (0.73, 12) or (0.95, 10.5), with α_o = 1.1 for $N_{c,circle}$ = 9.5. Jeanjean et al. (2006) performed double-walled pull-out tests, allowing separation of F_o, F_i and Q_{base} components, on D = 1.88 m, L = 11.25 m (L/D = 6) sealed caissons in NC kaolin clay. Combinations of α_o and $N_{c,circle}$ values of (0.85, 9.0) or (0.6, 12) were possible.

At the time of writing, no confirmation has been found from high-quality instrumented large-scale axial pile load tests in clay. This probably stems from the fact that piles in clay (usually L/D > 10) are friction piles, for which parameter α_o is far more important than $N_{c,circle}$. For intermediate "plugged" foundations (0.5 ≤ L/D ≤ 10), both α_o and $N_{c,circle}$ are relevant.

10.5.5.2 Skin friction and end-bearing

Another geotechnical myth is that skin friction and end-bearing contributions to pile axial capacity are uncoupled (e.g. Section A.8.1.4.2.1, ISO 19901-4:2016). In fact, skin friction and end-bearing are weakly coupled.

Load spread of skin friction increases surcharge (and hence end-bearing) at foundation tip level. This coupling is more apparent in sands than clays, especially in the other direction – end-bearing pressure increases normal stress/skin friction at/near tip level. This fact is reflected in CPT q_c-based methods for assessing pile capacity in sand.

10.6 MAXIMUM LATERAL RESISTANCE – SUPPORT FOUNDATIONS

VHM loads are at seafloor (see Figure 3.1a). Care should be taken in distinguishing between free and fixed head resistances H: "Fixed head" H_{max} occurs when the foundation cannot rotate (rotation $\theta_{xz} = 0$) and V = 0. "Free head" H_o is when V and M are both zero. Figure 10.3a shows their positions in the MH plane. Normalised H_{max} and H_o values versus foundation embedment ratio for various undrained shear strength profiles (Figure 10.3b) are given in Figure 10.3c and 10.3d. For embedment ratios L/D > 3, reasonable N_p values are 10.5 and around 3 for most clay profiles, and hence the ratio $H_{max}/H_o \approx 3.5$.

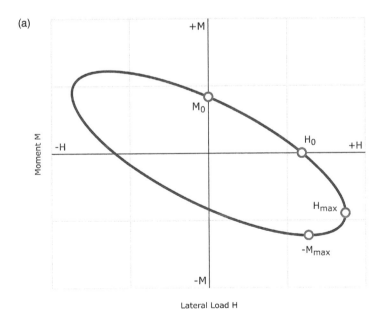

Figure 10.3 Maximum lateral resistance in clay. Normalised resistance factor N_p = $H/(LDs_{u,av,L})$ (a) MH plane, lateral resistances "fixed head" H_{max} and "free head" H_o (Kay and Palix, 2010).

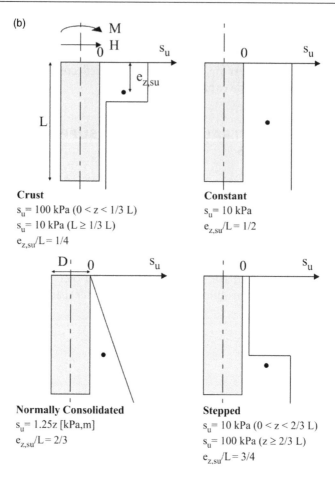

Figure 10.3 Continued: Maximum lateral resistance in clay. Normalised resistance factor $N_p = H/(LDs_{u,av,L})$ (b) Soil profiles.

Randolph et al. (1998) used AGSPANC (Advanced Geomechanics, 2002), an upper bound solution software, which can give optimistic values at low embedment ratios (Andersen et al., 2005). Kay (2015) used FEA and ignored internal scooping (if any) at low embedment ratios, where it can be seen that H_{max} and H_o values tend to converge. This is a consequence of lateral resistance being dominated by base shear, not lateral pressure. At higher embedment ratios, H_{max} values for uniform soil become constant, and both solutions agree with one another and also with the Randolph and Houlsby (1984) solution for a laterally loaded disc.

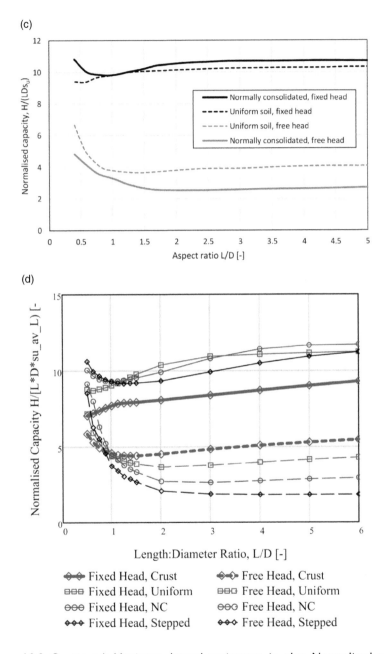

Figure 10.3 Continued: Maximum lateral resistance in clay. Normalised resistance factor N_p = $H/(LDs_{u,av,L})$ (c) Horizontal capacity of suction caissons (Supachawarote, 2006), N_p versus L/D (α_o = 0.65) (Kay, 2015).

10.7 MAXIMUM LATERAL RESISTANCE AND LUG POSITION (SUCTION ANCHOR PILES)

These paragraphs are about suction anchor pile lateral capacity. These are essentially rigid caissons. Analyses of laterally loaded *piles* are not discussed. The terms anchor chain and anchor line are synonymous.

Lateral performance, particularly moment failure of the steel at design loads, may be governing (rather than soil failure). Providing additional steel stiffeners at or near the lug level should prevent this. Note that additional steel increases installation resistance.

Unless load and chain inclination is provided at lug level, a separate analysis is required of the anchor chain below seafloor. This is to assess the decreased load and increased uplift angle, θ_{lug}, at lug level (see Figure 3.1b). Note that competent soils, such as sand and stiff clay, cause significant changes to the chain profile, particularly a further θ_{lug} increase. In such cases, the lug level is usually restricted to relatively shallow depths below seafloor in order to avoid axial pull-out. Equilibrium equations for an embedded anchor chain have been given by Degenkamp and Dutta (1989) and Neubecker and Randolph (1995).

If there is no restriction about lug depth, then maximum horizontal resistance (H_{max}) is obtained such that, in the ultimate limit state, the caisson behaves as a "fixed head" body, translating laterally with zero rotation as shown on Figure 8.1b. In this case, optimum lug level (on the caisson shaft) is determined by the intersection of the line of action of the lug chain load and the caisson centreline. In practice, the lug level is just below optimal (say, around 0.5 m for a 3 m diameter caisson). This ensures backwards rotation at failure, reducing potential for a tensile crack to open on the caisson trailing edge.

The optimum lug level is close to the weighted average of the shear strength profile, and the general expression for any shear strength profile is:

$$e_z = \int_0^L s_{u,z}\, z\, dz\, /\, \int_0^L s_{u,z}\, dz \qquad (10.13)$$

Values of e_z are approximately 2/3 L for normally consolidated clay 0.5 L for constant shear strength clays, and 0.75 L and 0.25 L for the extreme "stepped" and "crust" shear strength profiles shown on Figure 10.3b. Small corrections are necessary for small embedment ratios to account for base shear (Kay and Palix, 2011; Kay, 2015).

"Free head" and "fixed head" lateral resistances H_{max} and H_o are given by the following:

$$H_{max} = N_{p,fixed} L D s_{u,av,L}$$
$$H_o = N_{p,free} L D s_{u,av,L} \qquad (10.14)$$

where

L   = caisson embedded length
D   = caisson outer diameter
$s_{u,av,L}$ = average $s_{u,z}$ between caisson head (seafloor) and caisson tip (L)

and

$N_{p,fixed}$ = "fixed head" lateral bearing capacity factor
$N_{p,free}$ = "free head" lateral bearing capacity factor

Lateral bearing capacity factors N_p vary with foundation embedment ratio L/D; see Figure 10.3. For L/D > 1.5, reasonable "standard" N_p values are 10.5 and around 3 for most clay profiles for fixed and free head respectively. At shallower depths, they both increase and vary rapidly, due to the increasing base shear contribution.

From the generic equations $H = N_p \, L \, D \, s_{u,av,L}$, and $N_{p,fixed}/N_{p,free} = 3.5$, it is easily shown that, for NC clay, if the lug is moved upwards from optimum to (near) seafloor for a given diameter D, the embedded length L has to be multiplied by a factor $\sqrt{3.5} \approx 1.9$ in order to obtain the same H value. Similarly, if the embedment ratio (L/D) is kept constant, then the L multiplier is $\sqrt[3]{3.5} \approx 1.5$. Neither is geotechnically efficient. This underlines the necessity (if possible) to place the lug at/near optimum depth.

If the lug level is at seafloor, then lateral resistance is reduced by about a factor 3.5 because the caisson is able to rotate. The corresponding lateral resistance is the "free head" resistance (H_o).

If the anchor lug level is close to (just below) optimum, then a modest resistance decrease is warranted; see Figure 10.4. If more precision is required, then consider using one of the Table 3.1 analysis methods. Note incidentally that Figure 10.4 recovers the aforementioned "standard" N_p values (i.e. 10.5 and 3).

If anchor lug level is close to seafloor but not at optimum, then this is an intermediate case. Choices include

(a) assume "free" head resistance – slightly cautious
(b) use one of the Table 3.1 analysis tools.

Example – Fixed Head Pile. Use the same data as used previously in V_{max} example (Section 10.5.1), basically a 5 m diameter anchor pile to withstand 7500 kN factored load at lug level (T_{lug}) acting at 30° (θ_{lug}). Additional data included:

lug offset at pile outer radius (D/2, i.e. 2.5 m)
"fixed" head pile – i.e. lug at optimum depth $z_{lug,opt,fix}$

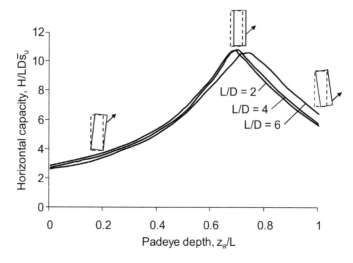

Figure 10.4 Effect of anchor lug level on lateral resistance (Randolph and Gourvenec, 2011).

Program CAISSON_VHM (Kay, 2015) results for the three "plugged", "coring" and "leaking" failure models are given in Table 10.4. This table shows that (for the aforementioned data):

- there is a progressive L increase – L values are $\approx$ 12.13 m, 14.10 m and 14.14 m for the "plugged", "coring" and "leaking" failure models respectively
- "plugged" is the most efficient – soil and drainage conditions permitting
- increasing L by around 2 m covers the possibility of "coring" and "leaking"
- all L/D ratios are around 2.5, and hence amenable to suction installation
- $z_{lug,opt,fix}/L$ (lug level depths : embedded length ratios) around 0.5–0.6 are less than the "2/3 L" value; this is due to the small embedment depth ratios (L/D $\approx$ 2.5), where base shear plays a role.
- since they are "stubby" foundations (L/D $\approx$ 2.5), smaller D values may well prove to be a more economical solution.

Note that these L values are for combined VHM load. Putting the lug level at seafloor, under pure V load, gives significantly higher L values – between 15.6 m and 22.3 m (see the Example in Section 10.5).

Figure 10.5 shows the corresponding three V_{lug}-H_{lug} envelopes, together with the actual T_{lug} load components, namely $V_{lug} = T_{lug} \sin(\theta_{lug})$ and $H_{lug} = T_{lug} \cos(\theta_{lug})$. The three shapes are dissimilar to V-H_{max} diagrams – see

Table 10.4 Foundation Embedded Lengths L for 5 m Diameter Anchor. 7500 kN Factored Tensile T_{lug} acting at θ_{lug} = 30° in Undrained Soil.

Parameter		V_{max} Failure Mode		
		"plugged"	*"coring"*	*"leaking"*
$L_{opt,fix}$	[m]	12.13	14.10	14.14
$L_{opt,fix}/D$	[–]	2.43	2.82	2.83
$z_{lug,opt,fix}$	[m]	6.76	8.11	8.14
$z_{lug,opt,fix}/L$	[–]	0.56	0.58	0.58
Resistances + loads				
V_{max}	[kN]	5235	3601	1821
H_{max}	[kN]	6954	9442	9508
$B_{w,caisson}$	[kN]	383	435	436
$B_{w,plug}$	[kN]	Not applicable	Not applicable	1633
$B_w = B_{w,caisson} + B_{w,plug}$	[kN]	383	435	2069

Notes:
$z_{lug,opt,fix}$: caisson optimum lug level depth (below seafloor)
$L_{opt,fix}$: caisson optimum embedded length ("fixed head" condition)

Figures 8.9 (ISO 19901-4) and 10.12 (Supachawarote et al., 2004). For example, taking the "plugged" failure model, (a) H_{lug} first increases, and then decreases, as θ_{lug} decreases from 30° to 0° and (b) V_{lug} first increases, and then decreases, as θ_{lug} increases from 30° to 90°. The latter is beneficial: T_{lug} exceeds 7500 kN should θ_{lug} exceed 30°.

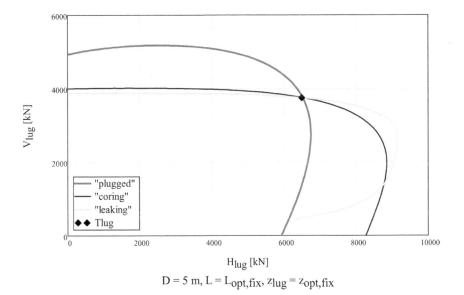

$D = 5$ m, $L = L_{opt,fix}$, $z_{lug} = z_{opt,fix}$

Figure 10.5 V_{lug}-H_{lug} envelopes and design T lug load.

The V_{lug}- H_{lug} envelope shape is the locus of points where V_{lug} and H_{lug} loads (at constant r_{lug}, z_{lug}, varying θ_{lug}) intersect the "tongue"-shaped VHM resistance envelope in 3D VHM space shown in Figure 2.2e. This does not have the same shape as ellipsoidal V-H_{max} diagrams, which are vertical slices through the same envelope at H_{max}.

10.8 MAXIMUM TORSIONAL RESISTANCE

In assessing maximum torsional resistance, due consideration should be given during site investigation and interpretation to the possible occurrence of discrete layers of low strength material along which displacements could preferentially occur. Failure due to the formation of internal mechanisms within the confined soil plug (above skirt tip level) should also be considered.

The maximum available torsional resistance, T_{max} is given by the lesser of the "plugged" and "coring" failure models, namely:

$$T_{max} = \min\left(T_{inner} + T_{outer}, T_{outer} + T_{base} \right) \tag{10.15}$$

For a uniform circular foundation with no protuberances in clay, this gives:

T_{inner} = torsion shear resistance of the inside of the foundation
  = $\alpha_i \, s_{u,av,L} \, L \, \pi \, D_i^2/2$
T_{outer} = torsion shear resistance of the outside of the foundation
  = $\alpha_o \, s_{u,av,L} \, L \, \pi \, D^2/2$
T_{base} = torsion shear resistance over the full foundation base
  = $s_{u,tip} \, \pi \, D^3/12$
α_i = clay-steel inner adhesion factor
α_o = clay-steel outer adhesion factor
$s_{u,av,L}$ = average s_u between seafloor and foundation tip (L)
$s_{u,tip}$ = soil undrained shear strength at foundation tip depth (L)
D = foundation outer diameter
D_i = foundation inside diameter
L = foundation embedded length.
Exterior protuberances (if any) provide a (small) additional torsional resistance ΔT, given by:

$$\Delta T = N_{c,prot} s_{u,prot} A_{prot} e_x \tag{10.16}$$

where

$N_{c,prot}$ = bearing capacity factor

$s_{u,prot}$ = undrained shear strength at protuberance depth

A_{prot} = protuberance bearing area

e_x = horizontal distance between neutral axis and lateral force resultant.

For "undrained" clay profiles, usually the "plugged" (not "coring" mode) is critical, and hence $T_{max} = T_{outer} + T_{base}$

10.9 TILT AND TWIST – ANCHOR PILES

These paragraphs are about intermediate anchor pile design for tilt (non-vertical or out-of-plumb) and twist (misalignment or misorientation). Both reduce holding capacity compared to a perfectly installed and orientated anchor. For resistance calculations, the most unfavourable situation (i.e. maximum tilt and maximum twist occurring at the same time) is often studied in detailed design. Typical specifications for installation tolerance are ±5.0° for pile tilt and ±7.5° for pile twist. Generally, maximum tilt (+5.0°) reduces capacity more than maximum twist (+7.5°).

Tilt is accounted for by increasing or decreasing the chain load angle at lug level (θ_{lug}) by the tilt angle. This approach is similar to the common assumption for inclined piles, namely that the axial and lateral responses are largely unaffected by small angles of inclination from the vertical (e.g. Poulos and Davis, 1980). Adding (e.g.) 5° tilt to θ_{lug} increases the V load component on the anchor, and subtracting 5° increases the H load component. Usually, anchors are more sensitive to V (than H) loading, and hence +5° tilt is more unfavourable than –5°. A non-horizontal seafloor, especially if it is sloping down in the same direction as the mooring line(s), should be considered. A reasonable procedure is to take the design tilt to be the greater of slope angle and tilt tolerance angle. For example, if the seafloor slope angle = 10°, and tilt tolerance = ±5.0°, then take design tilt = +10°.

Twist influences suction anchor foundation resistance less than tilt. This is seen in typical tolerance values: namely < ±5.0° for tilt and < ±7.5° for twist. Due to twist, part of the maximum available friction resistance will be used in torsional shear when the chain pulls on the suction anchor. The basic principles are given in Figure 10.6. Assuming the torsion load is fully applied first, a simple calculation can be done: knowing the applied torsion load and the maximum available resistance (on the suction anchor perimeter and base), the amount of shear utilised can be calculated. Vector analysis then gives the amount of shear available, and hence the reduction percentage, on the maximum frictional resistance for axial loading. To simplify design calculations, the pile-soil outer adhesion factor (α_o) for axial loading can therefore be decreased by the reduction percentage. Typically, this α_o decrease is small – around 5% (Senders and Kay, 2002). In the worked example that follows, the α_o decrease is around 2.5% (was 0.65, now 0.635). Interested

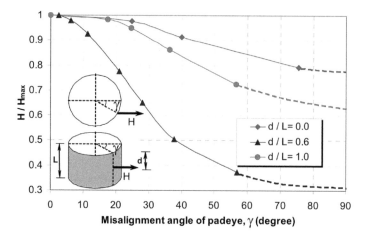

Figure 10.6 Effects of twist on suction anchor holding capacity (Taiebat and Carter, 2005).

engineers may care to repeat the example calculation for 15° (was 7.5°) twist. This occurred at Na Kika, Gulf of Mexico, details of which were given in Section 6.1.2. Large diameter foundations (D > 15 m, say) will probably give α_o decreases > 5%, and, if T_{load}/T_{max} is large, T_{load} – VHM resistance interaction effects (Section 10.10) may need to be considered.

10.9.1 Example – twist

To illustrate the difference twist (misalignment) makes to clay-steel outer adhesion factor α_o, consider a typical anchor pile in a NC soil profile. The relevant data are:

Pile: D = 5 m, L = 30 m
Soil: s_u = 2 + 1.5 z [kPa, m], α_o= 0.65
        ($s_{u,av,L}$ = 2 kPa + 1.5 kPa/m 15 m = 24.5 kPa, $s_{u,tip}$ = 2 kPa + 1.5 kPa/m 30 m = 47 kPa)
Lug: T_{lug} = 3 MN, θ_{lug} = 30°
Twist: ψ_{lug} = 7.5°

The calculations to find α_o^* (i.e. adhesion α_o allowing for torsion load T_{load}) are straightforward. They are as follows:

V_{lug} = pile axial load due to T_{lug}
     = $T_{lug} \sin(\theta_{lug})$ = 1500 kN
H_{lug} = pile lateral load due to T_{lug}
     = $T_{lug} \cos(\theta_{lug})$ = 2598 kN
T_{load} = pile torsion load due to twist (see Figure 10.6a)
     = H_{lug} D/2 $\sin(\psi_{lug})$ = 2598 kN 2.5 m sin(7.5°) = 848 kNm

T_{max} = maximum available torsional resistance, assume "coring" failure mode

$= T_{outer} + T_{base}$

$= \alpha_o\, s_{u,av,L}\, L\, \pi\, D^2/2 + s_{u,tip}\, \pi\, D^3/12$

$= (0.65 \times 24.5\ \text{kPa}\ 30\ \text{m}\ \pi\ (5\ \text{m})^2/2) + (47\ \text{kPa}\ \pi\ (5\ \text{m})^3/12)$

$= 18{,}761 + 1{,}538 = 20{,}299\ \text{kNm}$

η_T = pile-soil shear mobilisation factor due to torsion load T_{load}

$= T_{load}/T_{max} = 848\ \text{kNm}/20{,}299\ \text{kNm} = 0.042$

F_o = pile shaft friction resistance

$= \alpha_o\, s_{u,av,L}\, L\, \pi\, D$

$= 0.65 \times 24.5\ \text{kPa}\ 30\ \text{m}\ \pi\ 5\ \text{m} = 7504\ \text{kN}$

η_V = pile-soil shear mobilisation factor due to axial load V_{lug}, ignore REB

$= V_{lug}/F_o = 1500\ \text{kN}/7504\ \text{kN} = 0.200$

$\eta_V{}^*$ = pile-soil shear mobilisation factor due to combined V_{lug} and T_{load}; see Figure 10.6b

$= \sqrt{(\eta_V{}^2 - \eta_T{}^2)} = \sqrt{(0.200^2 - 0.042^2)} = \sqrt{(0.038)} = 0.196$

$\alpha_o{}^*$ = clay-steel outer adhesion factor allowing for torsion load T_{load}

$= \alpha_o\, (\eta_V{}^*/\eta_V) = 0.65\, (0.196/0.200) = 0.635.$

10.10 RESISTANCE UNDER COMBINED VHM(T) LOADS

Values of maximum vertical resistance V_{max} (Section 10.5), maximum lateral resistance H_{max} (Section 10.6), and maximum torsional resistance T_{max} (Section 10.8) are not achievable under combined VHMT loads. This is because, like shallow foundations and unlike deep (pile) foundations, coupling occurs:

- Torsion loads decrease intermediate foundation VHM resistance.
- HM loads decrease intermediate foundation V and T resistance.

For routine design, these two factors are considered using numerical analysis models, either based on limit equilibrium and plastic limit analysis (Section 10.11) or a resistance envelope approach (Section 10.12). The 3D FEA is generally not used for preliminary (and the majority of routine) geotechnical design of intermediate foundations.

For routine design, it is considered reasonable to assume that T_{load} affects only axial V (not lateral HM) resistance. Justification for this is given by Taiebat and Carter (2005) – as shown in Figure 10.7 – Saviano and Pisanò (2017) – shown in Figure 10.8 – and also Suroor and Hossain (2015), not shown.

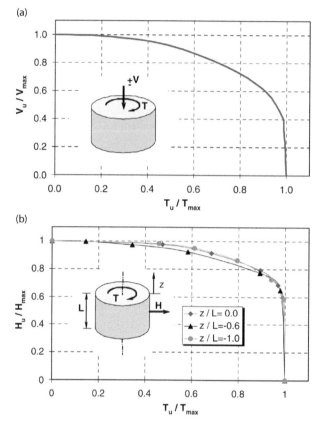

Figure 10.7 Effect of torsion (T) load on vertical (V) and horizontal (H) resistance (Taiebat and Carter, 2005).

10.11 IN-PLACE RESISTANCE ANALYSIS METHODS

10.11.1 General

Besides VHM(T) coupling, base shear and possible tension cracking need to be considered – this is as for shallow foundation design, but unlike pile foundation design. Unlike shallow foundations in clay, there is not generally a need to consider gap formation below the foundation base. Analysis methods include Limit Equilibrium, Plastic Limit and Resistance Envelope (aka Yield Function). The latter is a comparatively recent development in geotechnical engineering.

10.11.2 Undrained soil response (clay)

Support foundations for deepwater projects have many different mudmat and caisson types, each with various geometries, resulting in substantial numbers

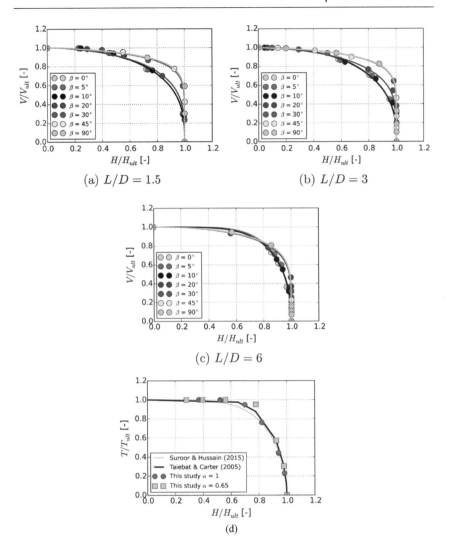

Figure 10.8 Effect of torsion (T) load on vertical (V) and horizontal (H) resistance (Saviano and Pisanò, 2017): (a, b, c) effects of misalignment on VH capacity envelope, (d) HT capacity diagram.

of preliminary optimisation analyses. Each analysis usually includes many (100+) VHM load cases, from which the governing ones have to be found. A key issue is to balance conflicting demands of speed, accuracy and cost.

Anchor foundations have fewer load cases and often less variation in geometry/aspect ratio. Capacity is sensitive to lug level depth and chain load angle, and key geotechnical issues are lug level and separate chain response analyses if T_{lug} loads are applied below seafloor. Complications include (a) iterating between separate caisson and chain models – which should use the same design soil profile – and (b) differing vertical and horizontal Factors of Safety on lug load T_{lug}.

Table 10.5 compares intermediate foundation resistance software functionality. Only CAISSON_VHM uses a yield function (details given in Section 10.12), and additional functionality includes:

- caisson tilt inclusion by adding tilt angle to θ_{lug} – see Section 10.9
- axial resistance V_{max} using either (a) "plugged", "coring" or "leaking"' failure models (Section 10.5.2) or (b) user-defined
- caisson buoyant weight B_w using either WT/D ratios or user-defined
- embedded chain profile using Degenkamp and Dutta (1989); these chain element equations were first stated by Vivatrat et al. (1982), and are identical to those given in Appendix A3 of DNV EP-R301 (2000)
- separate λT load factors on V and H components of lug load T_{lug}
- resistance modifications – see Section 10.12.7
- caisson twist (misalignment): reductions are usually small, less important than tilt (Senders and Kay, 2002), and can be simply done by decreasing α_o – see Section 10.9.

Table 10.5 Caisson In-place Resistance Software – Comparison Functionality (Kay, 2015).

Functionality	Software					
	A	B	C	D	E	F
Solution method	PL	ULS	ULS	YF	FE[2D]	FE[3D]
Support foundations	×	✓	✓	✓	✓	✓
Anchor foundations	✓	×	✓	✓	✓	✓
Chain profile	×	×	✓	✓	×	×
Rotational failure	✓	✓	✓	✓	✓	✓
Zero rotation failure	✓	×	✓	✓	✓	✓
Internal scoop failure	×	✓	✓	×	×	✓
Tension crack	✓	✓	✓	×	✓	✓
High axial load	×	×	✓	✓	✓	✓
Reduced REB	✓	×	×	✓	×	×
L, D optimisation	×	×	×	✓	×	×
Multiple L,D	×	×	×	✓	×	×
Multiple load cases	×	×	×	✓	×	×
Multiple soil layers	×	✓	✓	✓	✓	✓
Tilt	✓	×	✓	✓	✓	✓
Twist	×	×	✓	✓	✓	✓
Soil reactions	×	×	×	×	✓	✓

Notes:
A: AGSPANC (AG, 2003); FALL16 (OTRC, 2008); B: CANCAP2 (Fugro, 2009); C: CAISSON (Kennedy et al., 2013); D: CAISSON_VHM (Kay, 2015); E: BIFURC (Jostad, 1997); SPCalc (XG Geotools, 2014), F: ABAQUS, PLAXIS 3D etc.
ULS: limit equilibrium (neither LB nor UB); PL: plastic limit analysis (usually UB); YF: yield function (based on FE); FE: finite element analysis (slight UB), either 2D or 3D; LB: lower bound; UB: upper bound.

10.11.3 Drained soil response (sand)

There are no VHM(T) drained resistance envelope methods for intermediate foundations. In sand profiles, pile (rather than soil) failure occurs first. In addition, OWT monopile in-place response (e.g. tilt) is usually of major concern, and stubby support caissons are more likely to behave undrained than drained under extreme storm loads. The only method is that proposed by Broms (1964b); he used Plastic Limit Analysis (Upper Bound) for rigid piles under lateral (HM) load. V load coupling, base shear and internal scooping were not considered. Such situations require the use of 3D FEA.

10.12 VHM(T) RESISTANCE ENVELOPE METHODS

10.12.1 General

This section can be skipped by engineers uninterested in foundation optimisation. The majority of this section presents methods for undrained soils. This is because, as noted, a generic envelope for intermediate (and also for deeply embedded surface) foundations in drained sand under VHM loads is not yet available.

10.12.2 Undrained soil response ("clay")

Intermediate foundation capacity in clay may be quickly and reasonably assessed using VHMT resistance envelope theory. If failure envelopes and ultimate limit state (ULS) can be described by algebraic expressions, iteration can be employed to identify minimum foundation area for given soil strength and loading conditions.

For surface and shallow foundations (embedment ratio < 0.5), resistance envelopes are presented in section ISO 19901-4:2016 Clause 7.3.5 and Annex A.7.3.5. Because of their complexity and sensitivity to the soil s_u profile, only a few results are currently available. These are for a limited number of foundation shapes and undrained shear strength profiles, some of which include or exclude base tension, and several of which are described by approximating algebraic expressions enabling foundation optimisation (Gourvenec et al., 2017).

10.12.2.1 Resistance envelopes for undrained soil response ("clay")

Table 10.6 summarises resistance envelopes available for intermediate foundations in clay. They are in chronological order, earliest first. A common feature to all is that, since they are all FEA based, unlike ULS/plasticity based methods (e.g. AGSPANC/FALL16/CANCAP2/HVMCAP), one is "not getting back one's own input".

It can be seen that, for VHM loading, the most comprehensive work is that given by Kay and Palix (2010, 2011). Their resistance envelope is

Table 10.6 Resistance Envelopes for Intermediate Foundations in Undrained Soil (Clay).

Reference	Envelope	Soil s_u profile	L/D	Remarks
Supachawarote et al. (2004)	V-H_{max}	NC	1.5, 3 and 5	
Taiebat and Carter (2005)	VHT	Uniform	2 and 4	
Kay et al. (2010–2015)	VHM	NC, uniform, stepped and crust	0.5 - 6	CAISSON_VHM
Van Dijk (2015)	VHM	NC, uniform and stepped	0.125 - 2	Internal scoop
Sørlie (2013)	VHMT	NC	5	Project specific
Ahn et al. (2014)	V-H_{max}	NC and uniform	2 - 10	
Gerolymos et al. (2015)	VHM	Uniform	1, 2 and 3	Square caisson
Suroor and Hossain (2015)	VT	NC and uniform	1, 3 and 7	V&T applied at $z_{lug}/L = 0.7, \alpha = 1$

Note: NC, uniform, stepped and crust soil s_u profiles given on Figure 10.3b.

based on around 5500 quasi 3D (2D Fourier) finite element analyses for a large number of caisson L/D values ($0.5 \leq L/D \leq 6$) and soil s_u profiles (Figure 10.3b).

The resistance envelope is "tongue"-shaped (see Figure 10.9) in VHM space, and consists of a rotated ellipse in M-H space (Figure 10.9b), and recovers the familiar ellipsoidal shape of VH diagrams for shallow foundations in V-H_{max} space (Figure 10.9c). The complete envelope is described by three simple equations, and all parameter values are smooth functions of foundation embedment ratio (L/D) and non-dimensional soil shear strength profile ($e_{z,su}/L$). For a multi-layered soil profile, interpolation is used between the s_u profiles shown in Figure 10.3b. Major assumptions included no tension crack and no internal soil scoop failure. Torsion loading, usually small, can be accounted for by decreasing s_u (Senders and Kay, 2002; DNV, 2005). Benchmarks (including Andersen et al., 2005; Randolph and Houlsby, 1984; Taiebat and Carter, 2010), plus design optimisation examples for support and anchor foundations, are given by Kay and Palix (2010, 2011). Van Dijk (2015) gave resistance envelope equations for open-ended caissons, considering internal scooping.

The results can be applied to any type of rigid intermediate foundation – either support or anchor pile. For anchor foundations, since the VHM load reference point is on the centreline at seafloor (see Figure 3.1), and the foundation is essentially rigid, chain loads and inclination angle at lug level (T_{lug} and θ_{lug}) need to be transformed to seafloor VHM load using the three equations

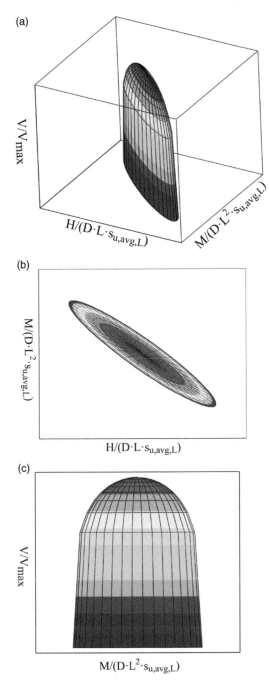

Figure 10.9 Intermediate foundation resistance envelope: (a) "tongue"-shaped VHM envelope, (b) rotated ellipse at base (the MH plane) and (c) an ellipsoid in the V-H max plane. LRP at seafloor (Kay, 2013).

$$V = V_{lug} \tag{10.17}$$

$$H = H_{lug} \tag{10.18}$$

$$M = -V_{lug}r_{.lug} - H_{lug}z_{lug} \tag{10.19}$$

where

V_{lug} = $T_{lug} \sin(\theta_{lug})$
H_{lug} = $T_{lug} \cos(\theta_{lug})$
r_{lug} = lug radial offset from foundation centreline
z_{lug} = lug level depth below seafloor

No anchor pile tilt (θ_{tilt} = 0) is assumed in the aforementioned equations, and (restoring) M values are therefore slightly cautious. This is because in the MH plane, the M load is reduced, meaning that the MH load data point lies further away from (and inside) the resistance envelope. These second order effects are small and are therefore usually neglected. If required, lug coordinates (r_{lug},z_{lug}) can be transformed for any θ_{tilt} (clockwise positive) using the equations:

$$r_{lug,\theta tilt} = \cos(\theta_{tilt})r_{lug} - \sin(\theta_{tilt})z_{lug} \tag{10.20}$$

$$z_{lug,\theta tilt} = \sin(\theta_{tilt})r_{lug} + \cos(\theta_{tilt})z_{lug} \tag{10.21}$$

10.12.2.2 Example seafloor VHM loads

Consider an anchor pile D = 5 m, with r_{lug} = D/2 = 2.5 m and z_{lug} = 16 m. Pile is subjected to load T_{lug} = 10 MN acting at an angle θ_{lug} = 30°. Find seafloor VHM loads for tilt angles θ_{tilt} = 0 and +5°.

For θ_{tilt} = 0, we have

V = $T_{lug} \sin(\theta_{lug})$ = 5000 kN
H = $T_{lug} \cos(\theta_{lug})$ = 8660 kN
M = $-V_{lug} r_{.lug} - H_{lug} z_{lug}$ = −5000 kN × 2.5 m − 8660 × 16 m = −151,060 kNm

For θ_{tilt} = +5°, the seafloor V and H loads remain unchanged, $\cos(\theta_{tilt})$ = 0.996 and $\sin(\theta_{tilt})$ = 0.087, and the transformed lug coordinates are given by

$r_{lug,\theta tilt}$ = $\cos(\theta_{tilt})r_{lug} - \sin(\theta_{tilt})z_{lug}$ = 0.996 × 2.5 − 0.087 × 16 = 1.096 m
   (was 2.5 m)
$z_{lug,\theta tilt}$ = $\sin(\theta_{tilt})r_{lug} + \cos(\theta_{tilt})z_{lug}$= 0.087 × 2.5 + 0.996 × 16 = and 16.157 m
   (was 16 m)
and

M = $-V_{lug} r_{lug,\theta tilt} - H_{lug} z_{lug,\theta tilt}$ = −5000 kN × 1.096 m − 8660 × 16.157 m =
   −145,400 kNm *(was −151,060 kNm, a 4% reduction)*

10.12.3 MH ellipses

For intermediate foundations, it is convenient to work in terms of dimensionless M and H values given by the equations (Figure 10.10)

$$H* = H / (LDs_{u,av,L})[-] \qquad (10.22)$$

$$M* = M / (L^2Ds_{u,av,L})[-] \qquad (10.23)$$

MH ellipses are defined by the following three parameters

(i) Φ_{MH} – rotation angle
(ii) a_{MH} – major semi-axis length
(iii) b_{MH} – minor semi-axis length

Ad (i): Φ_{MH} – rotation angle
M/H $\approx$ e_z/L (1/2 for uniform s_u profile, 2/3 for NC s_u profile, 3/4 for "stepped" s_u profile, see Figure 10.3

$$\Phi_{MH} \approx atan(e_z / L) \qquad (10.24)$$

Ad (ii): since H_{max} lies close to the semi-major axis (see Figure 10.11)

$$a_{MH} \approx cos(\Phi_{MH}) / H_{max}*; H_{max}* \approx 10.5 \qquad (10.25)$$

Ad (iii): since we have an almost straight line between H_o and M_o (see Figure 10.11)

$$b_{MH} \approx H_o * sin(\Phi_{MH}); H_o* \approx 3 \qquad (10.26)$$

Hence, one can obtain Φ_{MH}, a_{HM} and b_{HM} values from

(a) soil s_u profile
(b) caisson "free head" (H_o) capacity
(c) caisson "fixed head" (H_{max}) capacity

Figure 10.3c and 10.3d have compared normalised CAISSON_VHM free and fixed head resistance values with those of Randolph et al. (1998) obtained using AGSPANC (Advanced Geomechanics, 2002).

10.12.4 V-Hmax ellipsoids

Senders and Kay (2002), Supachawarote et al. (2004) and Kay and Palix (2010) all found that V-H_{max} resistance envelopes for caissons in clay (L/D in the range 0.5–6.0) were ellipsoidal and could be fitted with equations of the type

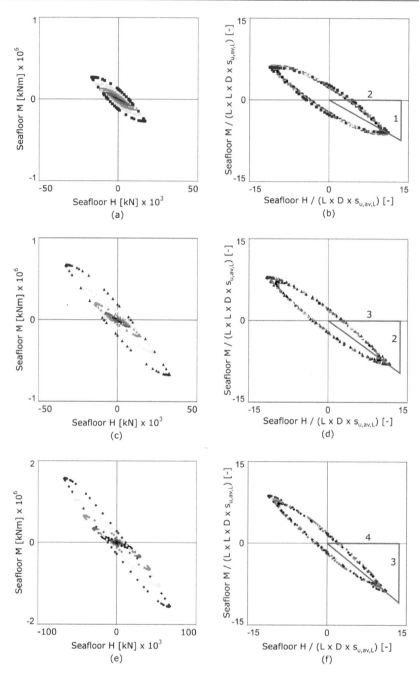

Figure 10.10 MH ellipses at V = 0. Summary final data points, caisson D = 5 m, L/D = 1.5, 2, 3, 4, 5 and 6. Soil s_u profiles: (a) and (b) Constant, e z, $s_u/L = 1/2$, (c) and (d) NC, $e_{z,su}/L = 2/3$ and (e) and (f) Stepped, $e_{z,su}/L = 3/4$. (Kay and Palix, 2010)

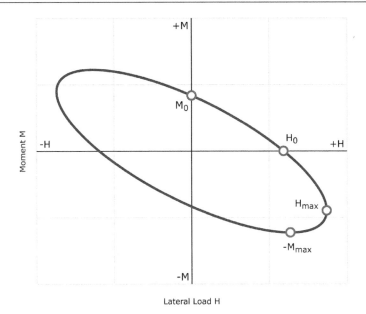

Figure 10.11 MH ellipse resistance envelope. Key data points H_0, M_0 and H_{max} (Kay and Palix, 2010).

$$\left(\left|\frac{H_{max,V}}{H_{max}}\right|\right)^{a_{VH}} + \left(\left|\frac{V}{V_{max}}\right|\right)^{b_{VH}} = 1 \qquad (10.27)$$

Figure 10.12 shows dimensionless V-H_{max} envelopes for different L/D values. As L/D increases, the less V-H interaction occurs, i.e. circles become more square. Hence, ellipsoid curvature parameters a_{VH} and b_{VH} are again functions of L/D and the s_u profile.

For example, for normally consolidated clay, Supachawarote et al. (2004) gives

$$a_{VH} = L/D + 0.5 \qquad (10.28)$$

$$b_{VH} = 4.5 + L/D/3 \qquad (10.29)$$

The high b_{VH} value implies that vertical resistance is less affected by horizontal load than vice versa.

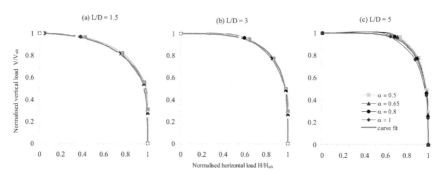

Figure 10.12 V-H$_{max}$ resistance. Dimensionless envelopes for differing caisson embedment ratios, L/D. Normally consolidated clay (Supachawarote et al., 2004).

10.12.5 VHM envelope – equations

The complete VHM resistance envelope is given by rotated ellipses in the MH plane (Section 10.12.3) plus ellipsoids in the V-H$_{max}$ plane (Section 10.12.4). Factor H$_{max,V}$/H$_{max}$ accounts for non-zero V load. The pair of parametric equations for a rotated ellipse are:

$$\frac{H_{load}\lambda_L}{LDs_{u,av,L}} = \frac{H_{max,V}}{H_{max}}\left(a_{MH}\cos(t)\cos(\phi_{MH})-b_{MH}\sin(t)\sin(\phi_{MH})\right) \quad (10.30)$$

$$\frac{M_{load}\lambda_L}{L^2Ds_{u,av,L}} = \frac{H_{max,V}}{H_{max}}\left(a_{MH}\cos(t)\sin(\phi_{MH})+b_{MH}\sin(t)\cos(\phi_{MH})\right) \quad (10.31)$$

and the ellipsoid equation is:

$$\left(\left|\frac{H_{max,V}\lambda_L}{H_{max}}\right|\right)^{a_{VH}}+\left(\left|\frac{V_{load}\lambda_L}{V_{max}}\right|\right)^{b_{VH}} = 1 \quad (10.32)$$

10.12.6 VHM envelope – yield function

The geometrical form of a rotated ellipse corresponding to the pair of parametric Equations (10. 30 and 10.31) is:

$$\left[\left(\frac{\cos(\phi_{MH})X+\sin(\phi_{MH})Y}{a_{MH}}\right)^2+\left(\frac{\sin(\phi_{MH})X+\cos(\phi_{MH})Y}{b_{MH}}\right)^2\right]$$
$$=\left|\left(\frac{H_{max\_V}}{H_{max}}\right)\right| \quad (10.33)$$

Eliminating ($H_{max,V}/H_{max}$) from Equations (10.30) and (10.31), the corresponding yield function, equation is obtained, namely:

$$f_{VHM} = \left(\frac{\cos(\phi_{MH})H* + \sin(\phi_{MH})M*}{a_{MH}}\right)^2$$

$$+ \left(\frac{-\sin(\phi_{MH})H* + \cos(\phi_{MH})M*}{b_{MH}}\right)^2 - \left[1 - \left(\left|\frac{V}{V_{max}}\right|\right)^{b_{VH}}\right]^{\frac{2}{a_{VH}}} \quad (10.34)$$

The first two terms of the yield function, Equation (10.34), are independent of V. They represent the geometrical form of a rotated ellipse in non-dimensional ($M*$, $H*$) space when $V = 0$. The third term, which is independent of H and M, has a non-dimensional axial load term, $|V/V_{max}|$, and double exponents. This term "shrinks" the yield surface (i.e. the $M*$, $H*$ ellipses) for non-zero axial load. For (almost) pure V load, especially when $|V/V_{max}| > 1$, numerical problems (imaginary numbers) with exponents are avoided by using a simpler form of Equation (10.34), namely

$$f_{vhm} = |V / V_{max}| - 1 \, \text{if} \, |V / V_{max}| > 1 \quad (10.35)$$

10.12.7 Modifying lateral and V_{max} resistance

Major *CAISSON_VHM* assumptions include (a) the caisson is fully surrounded by soil (b) suction is possible at the soil-steel interface (c) there is no failure within the soil plug (d) H and M load is co-planar.

In non-routine design, it may be necessary to consider exceptions such as:

(i) "gapping" in the active soil zone behind caisson for both support and anchor foundations (e.g. AG, 2003; DNV, 2005)
(ii) "anchor chain trenching" – a tapered wedge-shaped trench occurring in front of caisson due to anchor chain motions (e.g. Alderlieste et al., 2016; Colliat et al., 2018); this cannot occur for support foundations, as they are top loaded and have no chain
(iii) "vertical slot" above pad-eye for anchor foundations, which may occur during installation
(iv) "reduced end-bearing" for tension loaded anchor foundations (Clukey and Morrison, 1993; Clukey et al., 1995)
(v) "internal scoop" for support foundations at shallow L/D (Bransby and Yun, 2009; van Dijk, 2015)
(vi) "wings" for anchor foundations (Dührkop and Grabe, 2009)
(vii) non-co-planar HM load.

Apart from (vi), and possibly (vii), all decrease in-place resistance.

Since all geotechnical foundation models are inaccurate to a certain degree, and the CAISSON_VHM resistance model is no exception, it is reasonable to consider it to be also valid for these exceptions without a significant decrease in model accuracy.

Hence, four reduction factors (η) are available. Figure 10.13 shows how η_{aMH}, η_{bMH} and $\eta_{\varphi MH}$ modify the MH ellipse parameters a_{MH}, b_{MH} and φ_{MH}, and any η_{Fo} and η_{REB} combination modify V_{max}. Table 10.7 gives η examples/senses – values come from design specifications, literature or numerical analyses. Seafloor trenching reduces H_o more than H_{max} resistance, with a smaller V_{max} reduction which is mainly related to F_o (Alderlieste et al., 2016). Because of this, and the MH ellipse shape, η order is $\eta_{aMH} < \eta_{bMH} < \eta_{\varphi MH} < \eta_{Fo} < 1$. For "internal scoops" (possible only when L/D < 1.5), moment reduction is more severe than H_{max} (van Dijk, 2015). Hence η order is $\eta_{bMH} < \eta_{aMH} < 1 = \eta_{Fo} = \eta_{REB}$. Adding "wings" increases axial friction less than lateral resistance; hence η order is $1 = \eta_{REB} > \eta_{Fo} > \eta_{aMH} \approx \eta_{bMH}$. Note that, since a seafloor trench increases optimum lug level (OLL) depth, whereas wings have the opposite effect (OLL decreases), $\eta_{\varphi MH} > 1$ and $\eta_{\varphi MH} < 1$ respectively.

Lateral and axial resistances are modified by applying reduction factors to auto-computed a_{MH}, b_{MH} and V_{max} values. The shape of the yield surface (i.e. MH ellipses and V-H_{max} ellipsoids) is assumed unchanged, and the size of the yield surface is defined by parameters V_{max} and a_{MH}, b_{MH}.

Reduction factors η_{aMH} and η_{bMH} values < 1 may be used to decrease lateral resistance due to the possibility of a "tapered wedge"-shaped soil gap occurring in front of the caisson anchor foundation caused by anchor chain motions. Because optimum lug level (OLL) depth also increases, an $\eta_{\varphi MH}$ value > 1 is also required. To assess project-specific η_{aMH}, η_{bMH} and $\eta_{\varphi MH}$ values, 3D FEA are needed (Alderlieste et al., 2016).

The same principle is also applied to "gapping". Preliminary reduction factor η_{aMH} sources possibly include COFS (2003) API/Deepstar project and pairs of cases C1 and C3 and C2 and C4 from Andersen et al. (2005). From these, indicative reductions in V_{max} (giving η_{Fo} and/or η_{REB}) and H_{max} (η_{aMH}) may be assessed. A simpler (very cautious) approach is to simply reduce s_u by 50% over the crack depth and use η values = 1. Again, 3D FEA is necessary for detailed/final design.

Lateral resistance. MH ellipses: (a) $\eta_{aMH} < 1$, (b) $\eta_{aMH} < 1$, (c) η_{aMH} and $\eta_{aMH} < 1$, (d) $\eta_{\varphi MH} > 1$, (e) $\eta_{\varphi MH} < 1$
Ad (a) through (c):

- Use η_{aMH} and η_{bMH} to shrink major and/or minor axes, and hence lateral resistance.
- H_{max} resistance is proportional (directly related) to η_{aMH}.
- H_o resistance is a function of both η_{aMH} and η_{bMH}.

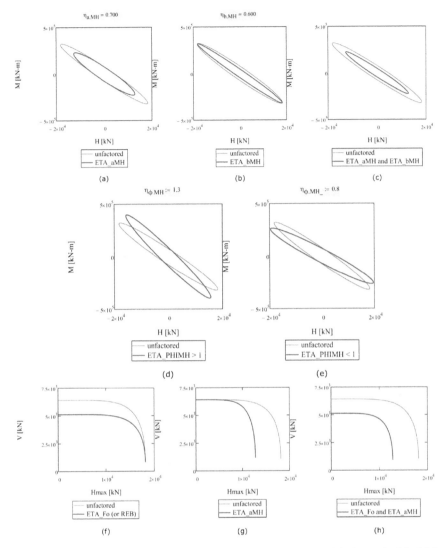

Figure 10.13 In-place resistance reduction. Top two rows (a) through (e): MH ellipses. Bottom row (f) through (h): V-H max ellipsoids.

Ad (d) and (e):

- Use $\eta_{\varphi MH} > 1$ to rotate ellipse clockwise (and hence increase OLL depth) and $\eta_{\varphi MH} < 1$ for anti-clockwise rotation (decrease OLL depth)

Ellipse rotation (i.e. $\eta_{\varphi MH} /= 1$) alters both H_{max} and H_o resistances; ellipse shape is unchanged.

Axial resistance. V-H_{max} ellipsoids (f) $\eta_{Fo} < 1$, (g) $\eta_{aMH} < 1$, (h) η_{Fo} and $\eta_{aMH} < 1$

Use either η_{Fo} and/or η_{REB} to reduce axial resistance V_{max}

Table 10.7 Resistance Reduction Examples.

Feature	Foundation type		Lateral			Axial	
	Support	Anchor	η_{aMH}	η_{bMH}	$\eta_{\varphi MH}$	η_{Fo}	η_{REB}
Gapping	×	✓	✓< 1	×		✓< 1	×
Seafloor trench	×	✓	✓< 1	✓< 1	✓> 1	✓< 1	×
Vertical slot	×	✓	×	×	×	✓< 1	×
Reduced REB	×	✓	×	×	×	×	✓< 1
Internal scoop	✓	✓	×	✓< 1	×	×	×
Wings	×	✓	✓> 1	×	✓< 1	✓> 1	×
HM⁺	✓	×	✓> 1	✓> 1	×	×	×
HM⁻	✓	×	✓< 1	✓< 1	×	×	×

Notes:

η_{aMH}	MH ellipse resistance semi-major axis length reduction factor
η_{bMH}	MH ellipse resistance semi-minor axis length reduction factor
$\eta_{\varphi MH}$	MH ellipse resistance rotation angle increase factor
η_{Fo}	caisson outer skin friction reduction factor
η_{REB}	caisson tip reverse end-bearing resistance reduction factor
REB	caisson tip reverse end-bearing resistance
HM	non co-planar HM load
+	overturning M load
−	restoring M load

10.12.8 Resistance comparisons

CAISSON_VHM was verified internally (Kay and Palix, 2010–2015). The following paragraphs give details of external (third-party) comparisons made with other programs.

10.12.8.1 Support foundations

Organisation A requested CAISSON_VHM analyses for three projects, comprising two support foundations and one anchor foundation, during their free trial period.

Organisation A made spot checks at L/D ≈ 0.5, 3 and 1, see L and z_{lug} *italicised* data in Table 10.8, and showed reasonable agreement. In addition, less than 1 s computer time was needed for all CAISSON_VHM analyses (3 runs, 1 or 2 load cases, 24 caisson diameters).

10.12.8.2 Anchor pile (and chain) foundations

As part of their acceptance criteria, Organisation B performed CAISSON_VHM–BIFURC–PLAXIS 3D comparisons for six anchor sites with T_{lug} up to ≈ 10 MN.

Table 10.8 Support Foundations – Geometry Comparisons.

ORGANISATION A		CAISSON_VHM	
Parameter	Organisation A Analysis	CASSION VHM Analysis	Remarks
Project 1 – SUBSEA TEMPLATE	Spreadsheet		Support foundation, free head. Single soil layer. Two VHM load cases. Optimum
Diameter D [m]	5.5	5.5	caisson embedded lengths
Embedded length L [m]	3	2.4	L_{opt} found for seven caisson diameters (2.5 m to 5.5 m).
Project 2 – SUBSEA STRUCTURE (MONOPILE)	Internal Software		Support foundation, free head. Two soil layers. Factored loads VHM(T) = 2.27, 2.0, 20.6 and 15.9 (MN,m).
Diameter D [m]	8.5	8.5	Reduced clay-steel outer
Embedded length L [m]	27.5	27.5	friction adhesion factor $\alpha_{o,twist}$ used to account for T load. Optimum caisson embedded lengths L_{opt} found for ten caisson diameters (4.0 m to 8.5 m).
Project 3 – SUCTION ANCHOR (MOORING SYSTEM) [1]	PLAXIS 3D		Anchor foundation, fixed head. Two soil layers. Factored chain load 17 MN applied at seafloor with an angle of 24°. Optimum
Diameter D [m]	12	12	caisson geometry
Embedded length L [m]	15	14	(embedded lengths $L_{opt,fix}$ and optimum lug level
Lug level depth, z_{lug} [m]	27.5	27.5	depths $z_{lug,opt,fix}$) found for seven caisson diameters (9 m to 15 m).

Note: PLAXIS 3D analysis was not optimised

CAISSON_VHM (Kay, 2015) uses a yield function based on ≈ 5500 quasi-3D FEA analyses. BIFURC (NGI, 1997) is a 2D FEA plane strain program taking 3D effects into consideration using side shear roughness factors. It uses 8-noded isoparametric elements (pile and soil mass) and 6-noded interfaces (pile-soil). PLAXIS 3D (PLAXIS, 2013) is a 3D FEA program. Both BIFURC and PLAXIS 3D are non-linear. As might be expected, the corresponding CAISSON_VHM–BIFURC–PLAXIS 3D computer run times are in the approximate ratio 1 : 100 : 1000.

Based on the results given in Figure 10.14, it is seen that BIFURC predicts the lowest resistance, and the PLAXIS 3D results, where available, provide the highest resistance. CAISSON_VHM provides the mid-prediction.

10.12.9 MH ellipse – design examples

Simple calculations can be made without having to use the yield function/ full resistance envelope equations. As an example, find embedded lengths L for a 5 m diameter caisson, when subjected to three load cases (overturning moment, zero moment and restoring moment). The soil is normally consolidated soft clay $(s_u(kPa) = 2 + 1.5z(m))$. Assume that V load is small and take "standard" N_p values of 3 and 10.5 for the "free head" and "fixed head" situations respectively. Figure 10.15 shows the MathCAD calculations and associated results (L = 7.5 m, 5.5 m and 2.5 m). Note that the first calculation makes use of the fact that the resistance envelope is essentially a straight line between points H_o and M_o (see Section 10.12.3 and Figure 10.11).

In practice, even for preliminary design, the soil profile is usually more complicated: N_p values vary with L/D, and slightly different embedded lengths L would be obtained. In addition, for the "fixed head" caisson, the

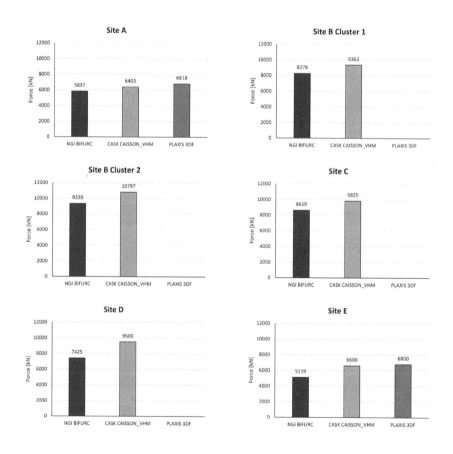

Figure 10.14 Comparison anchor pile T_{lug} – BIFURC, CAISSON_VHM and PLAXIS 3D.

extremely low embedment ratio L/D implies that cross plates may be necessary in order to prevent internal scoop failure. These calculations assume that the caisson is not heavily loaded axially – Figure 10.12 shows that V-H$_{max}$ ellipsoids are almost vertical until around V/V$_{max}$ = 0.5.

Figure 10.15 shows the example embedded length L calculations. 5 m diameter caisson in NC clay (s$_u$(kPa) = 2 + 1.5z) with factored lateral load H = 500 kN.

Case AA: overturning moment M = 1500 kNm (L = 7.5 m).
Case BB: free head, M = 0 (L = 5.5 m).
Case CC: fixed head, restoring moment (L = 2.5 m)

AA: CAISSON WITH OVERTURNING M

5 m diameter caisson in NC clay, su = 2 + 1.5z, factored loads H = 500 kN, M = 1500 kNm
V/V$_{ult}$ = small

work in normalised MH space

$M^* = M / L^2 D$ su_av
$H^* = H / L D$ su_av

Resistance: equation of straight line through Ho* with gradient m is

$$M_{star} = m(H_{star} - H_{o\_star})$$

where m is MH ellipse rotation angle (clockwise positive)

Loads: lie on a straight line through origin, equation is

$$\frac{M_{star}}{H_{star}} = \frac{M}{H \cdot L} \quad \text{or} \quad M_{star} = \frac{M \cdot H_{star}}{H \cdot L}$$

Hence, intersection (Resistance = Loads) is found by eliminating M* and solving for H*

$$m(H_{star} - N_{p\_free}) = \frac{M \cdot H_{star}}{H \cdot L}$$

$D := 5 \qquad M := 1500 \qquad H := 500$

assume $N_{p\_free} := 3 \qquad m := \frac{-2}{3}$

Guess value $L := 8$

Given

Three equations, three unknowns (L, H* and su_av)

$$m \cdot (H_{star} - N_{p\_free}) = \frac{M \cdot H_{star}}{H \cdot L} \qquad H_{star} = \frac{H}{(L \cdot D \cdot su\_av\_L)}$$

$$su\_av\_L = 2 + 1.5 \cdot \frac{L}{2}$$

$$\left(L \; H_{star} \; su\_av\_L \right) := Find\left(L, H_{star}, su\_av\_L \right)$$

$$\left(L \; H_{star} \; su\_av\_L \right) = (7.26 \; 1.85 \; 7.44)$$

$$H_{check} := H_{star} \cdot L \cdot D \cdot su\_av\_L$$

$L = 7.26 \qquad H_{check} = 500$

BB: FREE HEAD CAISSON - LUG LEVEL AT SEAFLOOR

Since M = 0 we have that

$$H = N_{p\_free} \cdot D \cdot L \cdot su\_av\_L$$

or

$$H = N_{p\_free} \cdot D \cdot L \cdot \left(2 + 1.5 \cdot \frac{L}{2} \right)$$

Hence embedded length L is given by

$$L := root\left[H - N_{p\_free} \cdot D \cdot L \cdot \left(2 + 1.5 \cdot \frac{L}{2} \right), L, 0, 10 \right]$$

$L = 5.47$

CC: FIXED HEAD CAISSON - LUG LEVEL AT OPTIMUM DEPTH

This is the same as BB but use Np_fix instead of Np_free

Assume

$$N_{p\_fix} := 10.5$$

$$L := root\left[H - N_{p\_fix} \cdot D \cdot L \cdot \left(2 + 1.5 \cdot \frac{L}{2} \right), L, 0, 10 \right]$$

$L = 2.47$

Note shallow caisson embedment ratio λ,
Thus Np factor no longer valid. $\lambda := \frac{L}{D} = 0.49$

Figure 10.15 Example embedded length L calculations. 5 m diameter caisson in NC clay (s$_u$kPa = 2 + 1.5z) with factored lateral load H = 500 kN. Case AA: overturning moment M = 1500 kNm (L = 7.5 m). Case BB: free head, M = 0 (L = 5.5 m). Case CC: fixed head, restoring moment (L = 2.5 m).

10.12.10 V-Hmax ellipsoid – design example

An example for optimum lug level (OLL) for anchors is given in Figure 10.16. This compares HARMONY (Griffiths, 1985, basis of CAISSON_VHM) with the Andersen et al. (2005) L/D = 1 and L/D = 5 benchmarks.

Unlike in the previous section, the equations are considerably more complex to implement in a simple MathCAD document for illustrative purposes. These challenges are because:

- there are two equations to be solved – one for the resistance envelope and another for the chain load at lug level – as summarised in Figure 10.17
- expressions for maximum caisson axial and lateral resistance, plus 3 MH ellipse constants (a_{MH}, b_{MH} and Φ_{MH}) and two VH ellipsoid constants (a_{VH} and b_{VH}) all have to be programmed in terms of caisson geometry (and shear strength profile).

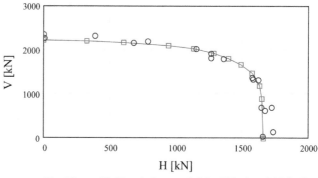

V - Hmax [L/D = 1.5, su = 1.25z, Wsub = 300 kN]

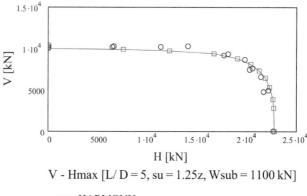

V - Hmax [L/ D = 5, su = 1.25z, Wsub = 1100 kN]

⊟⊟⊟ HARMONY
ooo ABAQUS and BIFURC (Andersen et al. 2005)

Figure 10.16 V_{max}-H_{max} ellipsoid example, optimum lug level (after Kay and Palix, 2010).

$$\left[\left(\left|\frac{H_{max\_V}}{H_{max}}\right|\right)^{a_{VH}} + \left(\left|\frac{V_{load}}{V_{max}}\right|\right)^{b_{VH}}\right] \cdot -1 = 0$$

$$\frac{M_{load}}{H_{load}} - ez\_H_{max} = 0$$

Figure 10.17 Optimum lug level - minimisation function equations.

10.12.11 VHM envelope – support foundation design example

Kay and Palix (2011) presented a typical design application for a 200 mT manifold founded on a caisson in normally consolidated clay (s_u profile given in Figure 10.18a) in the Mya North Field, Bay of Bengal, offshore Myanmar. The water depth at the manifold location is approximately 190 m. Using CAISSON_VHM, firstly, the 100+ load cases (operating, hydrotest, accidental and seismic) were first reduced in number by plotting factored HM load case data points together with HM resistance envelopes for various caisson geometries (Figure 10.18b). The most critical load cases are those lying closest to a resistance envelope; if the loads plot within the resistance envelope, then the design is satisfactory (loads < resistance) but not optimal (loads = resistance). It is seen that the seismic load cases are more critical than either operating or accidental loading. In addition, caissons with D = 4 m, L = 12 m will not provide sufficient resistance – they need to be slightly longer. Similarly, 5 m diameter caissons can be shorter than 12.5 m. Finally, note that this is a classic case of overturning H and M loading – all data points lie in the first quadrant of the HM diagram.

Secondly, the factored loads for the most critical load case (factored VHM loads = 2795, 1079, 7267 [kN, m]) were then used, together with the three envelope equations, to find the unknown required optimum embedment lengths L_{opt} for various caisson diameters D, as shown on Figure 10.18(c). Results obtained using CANCAP2 limit equilibrium analysis (Fugro, 2009) are also given for comparison. CANCAP2 automatically searches for a failure surface giving the lowest safety factor; analysis options include "internal scoops" – i.e. surfaces passing within soil plug inside the caisson (an example is given in Palix et al., 2010). It is seen that CANCAP2 and CAISSON_ VHM curves provide similar results for L/D > 0.5 where no internal scoop is developed. Had a very shallow caisson option been selected, the use of internal cross plates could have been considered for two reasons: (a) reduce settlement magnitude, and (b) prevent "internal scoop" failure.

As expected, a 4 m diameter caisson needs to penetrate more than 12 m (L_{opt} = 12.8 m), and a 5 m one less than 12.5 m (L_{opt} = 11.4 m). Note that L/D values are compatible with theory (the VHM resistance envelope equations are valid only for L/D > 0.5), and suction-assisted installation is likely

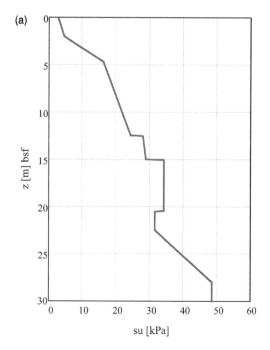

Figure 10.18 Caisson support foundation design example using CAISSON_VHM and CANCAP2: (a) soil s_u profile, NC clay, $s_u \approx 2z$; (b) MH plot-load cases and envelopes for given caisson geometries (V = 3000 kN); (c) required optimum caisson embedded lengths, L_{opt}, and embedment ratios, L_{opt}/D, for given caisson diameters D (factored VHM loads = 2795, 1079, 7267 [kN, m]) (Kay and Palix, 2011).

to give an insufficient FOS against base failure when L/D > 6). Based on the resulting L_{opt}-D curve, and the requirement to limit 30 year settlements to less than 0.25 m, the final caisson dimensions were D = 5.0 m, embedded length L = 11.4 m, total length = 12.4 m, side WT = 25.4 mm. The caisson weight was 73 tonnes.

PLAXIS 3D (PLAXIS, 2013) was used to verify the final caisson geometry for two of the most critical VHM load cases. The caisson was modelled as totally rigid with elastic material (steel). The clay-steel outer adhesion factor, α_o, was nominally 0.65, but with an appropriate reduction for torsion moment. The failure points (solid blue diamonds) obtained by PLAXIS 3D for the two load cases were added to the CANCAP2 MH resistance envelope and load case MH data (Figure 10.19). It can be seen that the resistance from the PLAXIS 3D FEA model is very slightly higher than the CANCAP2 resistance envelope. The PLAXIS 3D analysis needed 24 hours to run (2010) for each load case, whereas all the CAISSON_VHM resistance envelope analyses took less than 10 seconds.

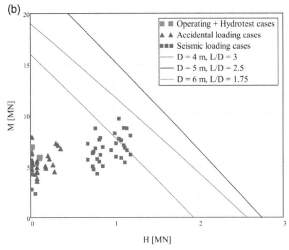

Factored MH Loads and MH Resistance Envelopes [V = 3 MN]

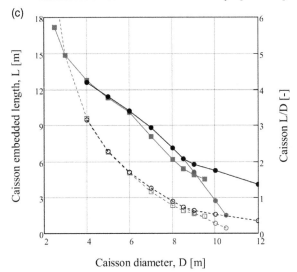

Caisson diameter, D [m]

■■■ L, VHM equations
●●● L, CANCAP2, no internal scoop
●●● L, CANCAP2, with internal scoop
⊟⊟⊟ L/D, VHM equations
⊙⊙⊙ L/D, CANCAP2, no internal scoop
⊙⊙⊙ L/D, CANCAP2, with internal scoop

Figure 10.18 Continued

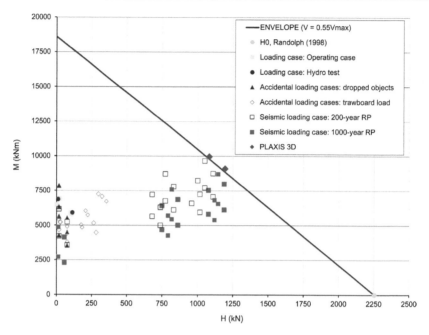

Figure 10.19 Caisson support foundation design example using CAISSON_VHM: PLAXIS 3D verification. Caisson D = 5 m, L = 11.4 m.

10.12.12 VHM envelope – anchor foundation design example

Kay (2015) presented a project example for FPSO anchors, offshore Nigeria. The objective was to find optimum foundation geometries for 0° and 5° tilt. Data were as follows:

caisson: D = 3 m, 3.5 m and 4 m. Wall and top plate thicknesses are D/200 and D/100

lug: on caisson outer face, z_{lug} at optimum depth

soil: $s_u = 2 + z$ $(0 \leq z \leq 5$ m), $s_u = 7 + 1.73$ $(z - 5)$ $(5$ m $\leq z \leq 20$ m) kPa, $\gamma_{sub} = 4$ kN/m^3

soil-pile: $\alpha_o = 0.6$ (including twist allowance), $\alpha_i = 0.65$.

pull-out V_{max}: lesser of plugged and unplugged conditions, $\eta_{REB} = 0.8$

chain: 81 mm diameter, submerged weight 1.12 kN/m.

loads and FOS: see Table 10.9.

Table 10.10 compares detailed results for both tilt angles. Figure 10.20a summarises the caisson geometry optimisation mode, which finds the optimum "fixed head" $z_{lug,opt,fix}$ and $L_{opt,fix}$ for given caisson diameter(s) D,

Table 10.9 FPSO Anchor Design Example: Chain Load Data.

Load Case	T_{z0} [kN]	θ_{z0} [deg]	λ_T [–]
1: All Lines Intact	2045	7.5	2.00
2: One Line Damaged	3377	13.7	1.50
3: Transient	4211	15.5	1.25

Note: 3.5 m OD, LC3

Table 10.10 FPSO Anchor Design Example. CAISSON_VHM Optimisation Results.

Parameter		0° tilt	5° tilt
$L_{opt,fix}$	[m]	13.2	13.8
$z_{lug,opt,fix}$	[m bsf]	8.5	8.7
T_{lug}	[kN]	4077	4097
θ_{lug}	[deg]	19.1	23.6
B_w	[kN]	192.2	199.3
V_{max}	[kN]	1934	2101
H_{max}	[kN]	5456	5981

Note: 3.5 m OD, Load Case 3

seafloor chain loads T_{z0} and inclinations θ_{z0} (i.e. T_{lug} and θ_{lug} at caisson lug level z_{lug}). A 5° tilt angle was used. Figure 10.20b shows that LC3 always has the largest embedment length L and is therefore critical. The associated maximum embedment ratio L/D is just over 6, suggesting that base heave installation may be an issue had D been less than 3 m. Figure 10.20c and Figure 10.20d provide optimum buoyant weight B_w and geotechnical efficiency (defined as $\eta = T_{lug}/B_w$) data used to finalise foundation geometry. LC3 efficiency η values decrease from ≈ 22 (D = 3 m) to 17 (D = 4 m), showing that "deeper is better". Final design was based on 3.5 m OD caissons (5° tilt, LC3), see Table 10.12. Additional checks subsequently made included capacity, lug level, tilt and twist. The sensitivity mode used (vary θ_{lug} and/or z_{lug}, find T_{lug}) is shown on Figure 10.21a. Figure 10.21b and 10.21c show T_{lug} sensitivity to θ_{lug} and z_{lug} respectively.

Less than 5 s computer time was needed for the two runs for Figures 10.20 and 10.21.

Alderlieste et al. (2016) give details of a more complicated anchor chain trenching example. This used PLAXIS 3D to first assess lateral resistance reduction factors, followed by CAISSON_VHM analyses for in-place resistance. Details are given in Section 12.12.

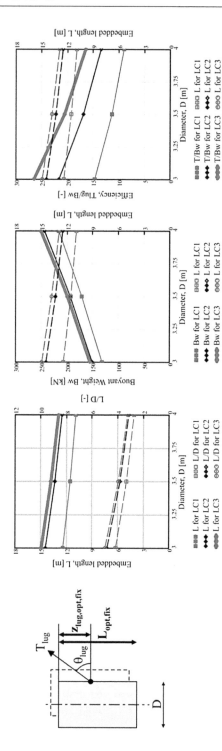

Figure 10.20 FPSO anchor design example using CAISSON_VHM, geometry optimisation mode (zero rotation). 5° tilt: (a) optimisation solution – (a) find $L_{opt,fix}$ and $z_{lug,opt,fix}$ for given D, T_{lug} and θ_{lug} (b) $L_{opt,fix}$, and L/D versus D, (c) B_w and $L_{opt,fix}$ versus D, (d) efficiency (T_{lug}/B_w) and L/D – D (Kay, 2015).

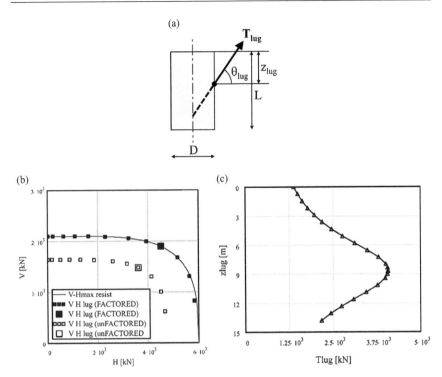

Figure 10.21 FPSO anchor design example, T_{lug} sensitivity mode (non-zero rotation). D = 3.5 m, LC3, λ_T = 1.25; (a) optimisation solution – find T_{lug} for given L, D, z_{lug} and θ_{lug}, (b) T_{lug} λ T and T_{lug} versus θ_{lug} at $z_{lug,opt,fix}$, (c) T_{lug} versus z_{lug} at θ_{lug} (Kay, 2015).

10.12.13 Drained soil response ("sand")

A generic envelope for intermediate (and deeply embedded surface) foundations under VHM loads is not yet available. Derivation is difficult: in the HM plane results are extremely sensitive to V load (unlike undrained clay). Also, there are more parameter variables (φ', Ψ', $\tan(\delta)$, K_o, etc.) and gapping to be considered, and at what caisson displacement to take failure resistance in numerical analyses. Finally, sand profile characterisation is more difficult than clay – it is rare to get "homogeneous" conditions (e.g. a constant relative density D_r) in the field.

Some work has been done by extrapolating the "rugby ball"-shaped envelope available for surface foundations (Butterfield and Gottardi, 1994) and VHM equation (Gottardi et al., 1999). Senders (2005), in order to provide tensile (and compressive) axial resistance, simply added an additional V term equal to outer friction and foundation steel submerged weight, although this severely underestimates the maximum lateral resistance due to extra embedment. Govoni et al. (2016) provide h_0, m_0 and a values for strips in 30° sand at L/D = 0 and 0.5.

10.13 RESISTANCE AT SHALLOW PENETRATION

Occasionally it may be necessary to assess intermediate foundation resistance at shallow penetration (embedment ratio L/D < 0.5). Examples include:

- limited penetration due to competent soil
- partial penetration and summer storm.

In such cases, shallow foundation design principles (e.g. ISO 19901-4:2016 Section 7.3.5 and Annex A.7.3.5) should be used. Scour is not generally considered.

10.14 RESISTANCE IN (WEAK) ROCK

10.14.1 General

The axial resistance and initial stiffness of weak rock will depend greatly on the mode of installation. Driving in weak rock will lead to the creation of a remoulded or crushed zone around the wall of the structure and potentially will lead to radial cracks around the structure (Section 9.13). The axial and lateral resistance of piles in weak rock are discussed in the following section for both installation cases.

10.14.2 Axial resistance of driven piles in weak rock

After driving, the rock around the pile wall will be partially or totally crushed. The material at the rock-pile interface will either be closer to granular material or cohesive material. In both cases, this crushing will result in decreasing normal stresses that will significantly reduce the shaft friction compared to a drilled and grouted interface. The axial friction in weak rock will mainly depend on rock mineralogy, void ratio, rock strength, rock mass properties and all factors affecting the in-situ stress (e.g. fractures and joints).

Two main approaches can be considered for assessing the shaft friction of driven piles in rock (Irvine et al., 2015):

- Effective stress methods (as for cohesionless soils):

$$Q_s = K.\sigma'_v \tan\delta.A_s < f_1.A_s \qquad (10.36)$$

where:

Q_s = shaft capacity

K = coefficient of lateral earth pressure

σ'_v = overburden stress

δ = friction angle between the rock and pile wall

A_s = surface area of the pile shaft

f_l = limiting skin friction.

The K factor should ideally be measured by pressuremeter testing that would account for in-place conditions such as rock mass stiffness and anisotropy.

- Total stress methods where a pile-rock adhesion is calculated (as for cohesive soils):

$$Q_s = \alpha.UCS.A_s < f_l.A_s \qquad (10.37)$$

Where:

α = adhesion factor

UCS = Unconfined Compressive Strength.

This latter method appears to be better suited to weathered mudstone and weak calcareous sedimentary rock. Terente et al. (2017) propose a lower bound adhesion factor formulation based on published driven piles data (see Figure 10.22).

The adhesion factor is given by the following equation:

$$\alpha = 0.11.(UCS)^{0.5} \qquad (10.38)$$

For cemented carbonate formations, ARGEMA (1994) propose an adhesion factor ranging between 0.1 and 0.2 based on Burt and Harris (1980) and Puech et al. (1988). Mention is made of field tests showing evidence of shaft friction up to 300 kPa for rock formation with UCS up to 5 MPa.

In chalk, driving will generate pore pressure in the remoulded material (also referred to as putty chalk) that will further reduce the effective stresses around the pile. CIRIA (2002) recommend considering a shaft friction just after driving of 120 kPa for high-density Grade A chalk and only 20 kPa for all other chalks. These recommendations do not account for effects of time and cyclic loading, both having opposite and significant impacts on chalk. Much effort is currently being made by the wind energy industry to better characterise the shaft friction in general or for specific projects. Extensive laboratory tests (see Houlston et al., 2017; Dührkop et al., 2017) or offshore or onshore pile tests campaigns (see Barbosa et al., 2015) are more and more commonly conducted to justify using higher friction values. A JIP project, ALPACA, was initiated in 2018 to propose new design methodologies for pile in chalk. This project includes a series of onshore pile tests

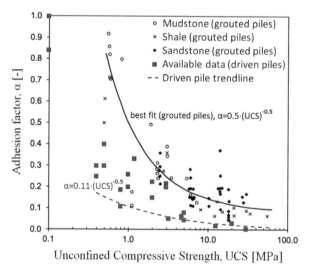

Figure 10.22 Adhesion factor (α) against rock material UCS (Terente et al., 2017).

conducted in the UK (St Nicolas site), comprising axial and lateral load tests (Jardine et al., 2019).

10.14.3 Axial resistance of drilled and grouted piles in rock

Pile construction factors are playing a great role in the quality of the grout/rock interface. For drilled and grouted piles, aspects that need to be considered are:

- transfer of steel/grout forces
- hole stability
- presence of granular material (risk of wash out)
- roughness of the hole
- cleaning of the hole (for both axial friction and end-bearing)
- fracturing breakdown (grout pressure)
- grout quality (strength, swelling/shrinkage, heat of hydration).

Hole stability is one of the first factors to consider for a drilled and grouted foundation. It is noted that when sub-layers of granular material are encountered in the rock matrix, the drilling fluid circulation will most probably wash out these materials. Wash out will produce large voids that could lead to stress relief in the rock mass, and also require a significant amount of additional grout to be injected to fill these voids. If pressurised, the injection of grout can also lead to amplification or creation of fractures.

Potential effects of drilling mud need to be considered when determining the shaft friction. The sidewall of the hole can develop a layer of slaked mud or clay, which will never gain the strength of the rock. Roughness of the hole wall is also an important factor. Depending on the roughness, the shearing will occur in a zone impacted by the drilling (slaked mud or crushed material) or within the rock mass. To ensure a good shear transfer, the roughness can be artificially increased by the use of an under-reamer at regular depths. Figure 10.23 shows the drilling head equipped with an under-reamer used for the Pluto project offshore Australia. Grooves were made every 1.5 m to perform so-called "ribbed" rock socket foundations for the Pluto jacket.

The cleanness of the drilled hole plays an important role on the end-bearing response. In general, the end-bearing capacity should be factored or neglected in the design, depending on pile construction aspects such as removal of drill cuttings from the base of the hole and presence of discontinuities within the rock mass.

Skin friction for socket piles can develop in one of three ways:

- through a shearing bond between the grout and the steel pile
- sliding friction between the grout shaft and the rock (when the drilled hole is smooth and grout infiltration limited)
- dilation of an unbounded rock-grout interface until asperities shear off.

For stubby piles submitted to a large number of cycles of loading, the axial resistance should not rely on the passive shear capacity between the steel and cast grout (c.f. DNVGL RP 0419 recommendations for grouted connections). Pile movement during grout curing could jeopardise the steel/grout bonding. In addition, cyclic lateral loading will tend to create a gap between the grout and the steel. The shear capacity of the interface can be increased

Figure 10.23 Pluto – drilling head (3m diameter) with underreamer. (Courtesy of Large Diameter Drilling (LDD))

by using shear keys, i.e. protuberances on the steel surface, on the interface that when subsequently cast into the grout provides a mechanical resistance to relative sliding between the steel and grout. API RP 2A WSD (2000) and ISO 19902 (2011) provide guidance on the number of shear keys to be considered.

In non-carbonate weak to moderately strong rock, high friction values have been recorded. Most published methodologies define the unit shaft friction of drilled and grouted piles as a function of the Unconfined Compressive Strength (UCS), independent of the type of friction developed (i.e. sliding or dilation). Table 10.11 lists the main methodologies proposed for shaft friction, f_s, where

$$f_s = \alpha.(UCS)^m \tag{10.39}$$

where:

UCS: Unconfined Compressive Strength

m = constant usually taken as 0.5

α: adhesion factor comprised between 0.2 and 0.45.

The large variability obtained by these empirical methods highlights the limitation by the use of the UCS as the only reference. Figure 10.24 shows the spread of adhesion factor for different rock types (Pells et al., 1980). In addition, these methods do not account for drilled hole dimensions (i.e. diameter and roughness), natural fractures, rock mineralogy and porosity.

Seidel and Collingwood (2001) proposed another approach incorporating the complex mechanisms of shear transfer at the interface between the socketed piles and the surrounding rock. They define a shaft resistance coefficient (SRC) incorporating the contribution of the rock strength (drained intact and residual strength parameters are used), socket roughness, rock mass modulus, Poisson's ratio and socket diameter. The formulation of the SRC is proposed as

Table 10.11 Ultimate Friction Formulation for Drilled and Grouted Piles in Weak to Moderately Strong Rock.

Drilled and grouted methods	Ultimate friction formulation
1. Rosenberg and Journeaux (1976)	$0.375(UCS)^{0.515}$
2. Horvath (1978)	$0.33(UCS)^{0.5}$
3. Horvath and Kenney (1979)	$0.2{-}0.25(UCS)^{0.5}$
4. Meigh and Wolski (1979)	$0.22(UCS)^{0.5}$
5. Williams and Pells (1981)	$\alpha{\cdot}\beta{\cdot}UCS$ (where β is a constant)
6. Rowe and Armitage (1987)	$0,45.(UCS)^{0.5}$

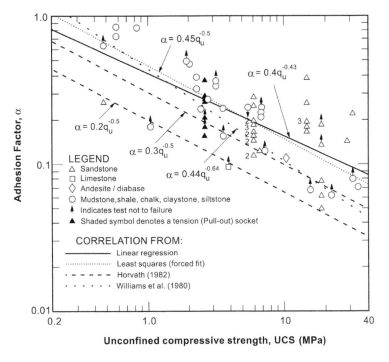

Figure 10.24 Shaft resistance correlations following Seidel and Collingwood (2001) for roughness classes R1-R3 of Pells et al. (1980).

$$SRC = \eta c \cdot \frac{n}{1+\upsilon} \cdot \frac{\Delta r}{d_s} \qquad (10.40)$$

where:

Δr is the mean roughness height

d_s is the socket diameter

ηc is the construction method reduction factor

n is the ratio E_m/UCS of rock mass modulus over unconfined compressive strength.

The effect of the SRC on the adhesion factor is provided in Figure 10.25. For slightly cemented carbonate rock (UCS < 5 MPa), ARGEMA (1994) suggest considering the ultimate shaft friction at the carbonate rock–grout interface proposed by Abbs and Needham (1985; see Figure 10.26). For chalk, CIRIA (2002) differentiate the axial resistance depending on the density. For low to medium density, the friction is a function of the average effective stress, $f_s = 0.8\ \sigma_v'$. For high-density chalk, the axial resistance is proportional to the UCS: $f_s = 0.1$ UCS.

The methods just outlined rely on the measurement of the unconfined compressive strength of the rock. More recent designs rely on an analogy between the displacement of the rock socket and displacement of small scale rock/grout interface loaded under constant normal stiffness (CNS) conditions. CNS conditions are believed to be more representative of the

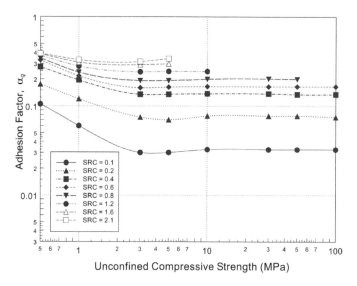

Figure 10.25 Effect of shaft resistance coefficient (SRC) on socket adhesion factor (Seidel and Collingwood, 2001).

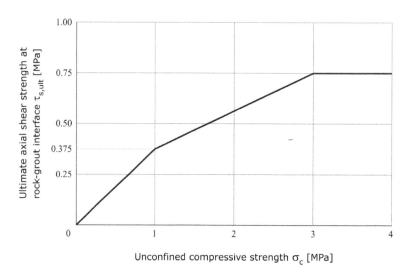

Figure 10.26 Ultimate shaft friction at the carbonate rock–grout interface (from Abbs and Needham, 1985).

evolution of normal stresses applied on the interface during rock socket pile displacement (Figure 10.27 and CFMS, 2019).

The imposed stiffness should be representative of the actual rigidity of the rock mass. The rock–grout interface is represented by a regular saw-tooth with a chord length and an asperity height corresponding to the roughness of the in situ hole. Guidance for characterising rock masses for designing drilled and grouted offshore pile foundations can be found in Puech and Quiterio-Mendoza (2019), and testing conditions are described by Stavropolou et al. (2019).

In the case of weak rock (i.e. grout being much stronger than the rock), the interface response will be governed by the rock behaviour. The failure surface will pass through the weak rock. For stronger rocks (i.e. strength similar to the grout), it is more the "geometrical" dilatancy that will govern the interface response.

The main challenge of this experimental methodology (based on CNS tests) is to get access to proper design parameters and installation conditions at an early stage in the design process. The measured shear force is very

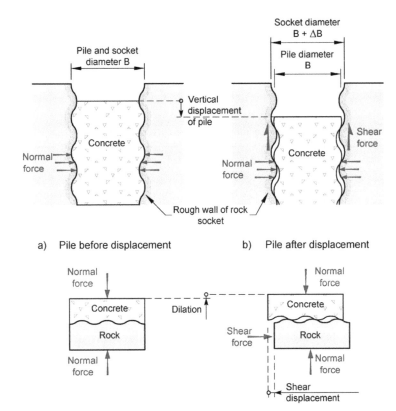

Figure 10.27 Idealised displacement of a rock socket.

sensitive to the shape of the asperities. It remains, however, the best way to understand the degradation of asperities under cyclic loading without requiring full-scale tests.

10.14.4 Lateral resistance in rock

It should be highlighted that experience of foundations installed in weak rock is relatively limited, particularly for driven piles. Rock is often encountered below upper soils and therefore usually mobilised far from its failure point.

The most commonly considered approaches to derive the lateral resistance of a pile in weak rock are listed in Table 10.12.

The first two p-y methodologies (Abbs, 1983; Fragio et al., 1985) are composed of an initial intact rock resistance followed by residual resistance at large strains, shown in Figure 10.28. In the Reese (1997) framework, the lateral pressure increases until reaching an ultimate value.

An alternative model, CHIPPER (Erbrich, 2004), takes cyclic loading into account explicitly. CHIPPER aims to capture the brittle stress-strain behaviour through an algorithm modelling the development of fractures ("chipping") near the surface of the rock (Figure 10.29).

Other methodologies for calcareous sand have been proposed (e.g. Novello, 1999; Dyson and Randolph, 2001) but are applicable only to weakly cemented calcareous sediments in which CPTs can penetrate.

Table 10.12 Methods to Determine Lateral Resistance in Weak Rock.

Reference	UCS (MPa)	Rock type	Failure mode	Based on
Abbs (1993)	0.5–5	Carbonated	Brittle near the surface and ductile deeper	Theoretical basis
Fragio et al. (1985)	9–36	Calcareous Claystone	Brittle near the surface and ductile deeper	Bored piles, D = 0.405 m
Reese (1997)	3–16	Various	Ductile	Bored piles, L/D = 6 to 11
Dyson and Randolph (2001)	CPT can penetrate	Cemented carbonated soil	Ductile	Centrifuge tests
Erbrich (2004)	No range provided	Cemented carbonated soil	Brittle near the surface and ductile deeper	Theoretical + centrifuge tests + 3D FE

Notes: L: embedment length, D: pile diameter, UCS: unconfined compressive strength

(a)

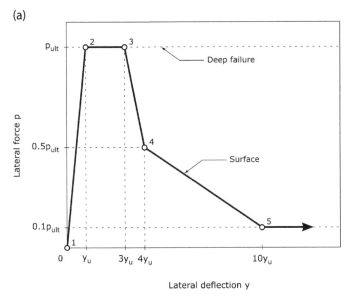

(b)

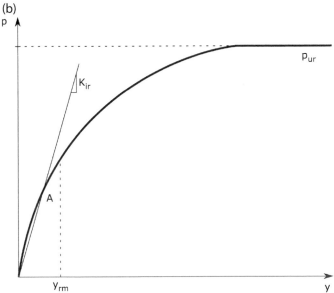

Figure 10.28 Characteristic shape of p-y curves for weak rock according to (a) Fragio et al. (1985) and (b) Reese (1997) methods. (With permission from ASCE)

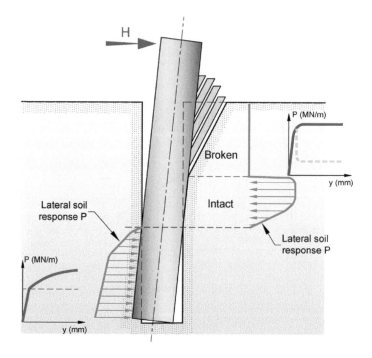

Figure 10.29 Concept of "chipping" and characteristic shape of p-y curves (after Erbrich, 2004).

The main limitations of these existing methodologies are as follows:

- Only a few field tests were considered.
- Most of the field tests included bored piles. The effect of pile installation by driving is therefore not considered explicitly (it could be by introducing a damage factor but with no guidance on how to define this damage factor).
- Methodologies were developed for slender piles (apart from the CHIPPER method also used for large diameter anchor piles, D = 5 m) and do not provide any recommendations for extrapolation to large diameter piles.
- The effect of cyclic loading is not discussed in these methods apart from the CHIPPER method proposing a cycle-by-cycle approach.
- These methods use UCS (or cone resistance for Dyson and Randolph, 2001) and Rock Quality Designation (RQD). They do not consider the porosity, carbonate content or fracture orientations.

10.15 GROUP RESISTANCE

10.15.1 General

Groups of suction foundations may be installed. Examples are given in Figure 10.30. They include two or more suction anchors for FPSO mooring points, and braced foundation clusters to ensure a near-vertical subsea template.

10.15.2 Considerations

Considerations include:

- centre-to-centre spacing – interaction potential
- braced or unbraced – kinematic constraints
- MH load – directionality

Like pile groups, the primary consideration is the spacing between individual foundations. At close spacings, when foundations interact, bracing (if any) imposes kinematic restraints – an unbraced self-installing platform behaves differently than a braced tripod.

Group load (capacity and response) is normally considered when the centre-to-centre spacing is < 8 D. Generally, for foundations in undrained "clay", the group capacity may be less than the single suction foundation capacity multiplied by the number of foundations in the group. Axial and lateral group deflections are normally larger than that of a single suction foundation subjected to the average load of the group.

10.15.3 Design procedures

There are no established procedures for designing groups of shallow or intermediate foundation systems.

Therefore, engineers resort to ad hoc (case-by-case) approaches.

10.15.4 Design tools

Design tools (listed in order simplest first) include:

- equivalent area foundation
- VH_{group} = sum(single); M_{group} = sum(V*lever arms)
- plastic limit analysis/resistance envelope
- 3D FEA.

10.15.5 Braced support groups

10.15.5.1 Shallow foundation groups

Various offshore structures are founded on multiple shallow foundations such that the kinematic constraint of the structure leads to the foundation

(a)

(b)

Figure 10.30 Examples of suction foundation groups: (a) four-legged structure (Courtesy Bo B. Randulff, Equinor), (b) cluster foundation (Courtesy SPT Offshore).

(c)

(d)

Figure 10.30 Continued: Examples of suction foundation groups: (c) cluster pile for Ceiba project (Courtesy SPT Offshore), (d) L6-B platform during sail-out (from Alderlieste and Van Blaaderen, 2015).

elements to act in concert. Examples include jackets permanently supported on suction cans, piled jackets temporarily supported by mudmats, and a variety of subsea infrastructure.

The vertical and horizontal capacity of structurally connected shallow foundation systems is relatively unaffected compared to a single foundation of equivalent bearing area and embedment ratio. However, the structural rigidity of a multi-footing foundation system enhances overturning capacity due to the structural connection between adjacent foundations (Murff, 1994; Fisher and Cathie, 2003; Gourvenec and Steinepreis, 2007; Gourvenec and Jensen, 2009). There is no established procedure for stability calculations of a shallow foundation system comprising co-joined footings, and ad-hoc approaches may involve simply summing the ultimate limit states of the individual footings under vertical and horizontal load and considering a push-pull mechanism for moment resistance, using a method of linear springs or carrying out project-specific finite element analysis (Randolph and Gourvenec, 2011).

10.15.5.2 Anchor pile groups

Cox et al. (1984) reported results of model laterally loaded pile groups in soft clay. Their data, summarised in Figure 10.31, suggests that groups of 3 and 5 side-by-side piles with centre-to-centre spacings of 3D to 4D (i.e. 2D-to-3D soil gap) develop lateral resistances approaching that of single isolated piles. The braced boundary condition is dissimilar to that usually encountered for pile anchor clusters (Figure 10.30).

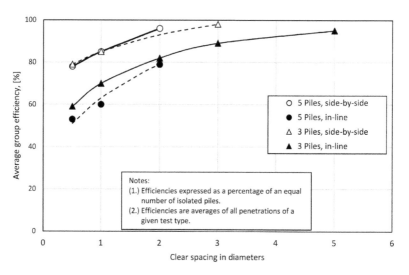

Figure 10.31 Laterally loaded anchor. Braced pile groups (Cox et al., 1984).

10.16 CLOSURE

This chapter has covered in-place resistance, both for support and anchor pile type intermediate foundations, and the following sections subdivide comments into those for "clay" and "sand".

10.16.1 Clay

As for installation, models for undrained "clay" are more developed than for drained "sand". Reasonably accurate models have been given to assess maximum available V_{max}, H_{max} and H_o ("fixed" and "free head") and T_{max} resistances. For anchor foundations, pull-out V_{max} has three possible failure models – "plugged", "coring" and "leaking" (see Figure 10.1), for which both applicability criteria and equations are provided.

VHM(T) interaction effects are important: HM load affects V resistance, and, to a lesser extent, T load affects V resistance. Like shallow foundations, VHM(T) yield envelopes for undrained soil are a promising development for preliminary/routine design. Equations and software are available. PLET and FPSO anchor design examples are also given.

Both FEA and centrifuge tests suggest that the original (Skempton, 1951) bearing capacity factor $N_{c,circle}$ is cautious. Hence Q_{tip} is probably underestimated. This has implications for both reverse end-bearing (in-place, this chapter) and plug heave (installation, Chapter 9) failure.

For anchor piles, V_{max}-H_{max} and V_{lug}-H_{lug}, envelope shapes differ; compare Figure 10.16 with Figure 10.5. This should be considered during detailed design when making sensitivity analyses.

10.16.2 Sand

Models for "sand" are less mature than for "clay". This is as expected: effective stress governs strength, and pore water build-up/dissipation is an additional complicating factor.

For sustained load, sands behave drained, and calculations can be easily made to assess both the compressive resistance V_{max} (which is typically very large) and the tensile pull-out resistance (which is very small). There is also a good understanding of near-surface foundations to combined VHM loads, largely based on experimental data. Yield surfaces defining possible VHM load combinations have been published, but these are for either shallow foundations or jack-up spudcans. Hence, as a start, a nice research project would be to derive a reasonably cautious model for assessing intermediate foundation resistance in drained sand under static VHM loading. Besides foundation geometry, which must take account of embedment, key soil parameters include friction angle φ' and dilatancy parameter ψ'. This research would hopefully lead to a drained resistance envelope and yield function similar to, but far more complex than, that developed for intermediate foundations in undrained soil.

For short-term loading (e.g. a single wave load or boat impact), resistance estimates are challenging because the equivalent undrained sand strength is very difficult to assess. Undrained strengths are principally determined by dilatancy parameter D (DNV, 1992) and the onset of pore fluid cavitation. Hence both vertical compressive and tensile resistances are typically very large, but difficult to assess. The resistance to combined moment and horizontal loads is also considered to be large. Like clays, VHM interaction effects must be considered. Resistance assessments become even more complex for cyclic load. An even more challenging research project would be to initially assess resistance in *undrained* sand under *static* VHM loading.

Chapter 11

In-place response

11.1 INTRODUCTION

This chapter deals with intermediate foundation response under static (Section 11.2), dynamic and non-dynamic cyclic loads (Section 11.3). It is seen that the response topics considered are basically similar to those for shallow foundation design.

The definition of dynamic and non-dynamic cyclic loads is as follows: a cyclic load can be termed dynamic if the rate of loading is such that inertial forces are significant compared to the static loads (e.g. boat impact, earthquake or ice shock forces). If inertial forces are negligible (e.g. wave forces), then the loading is said to be non-dynamic cyclic (Chan and Hannah, 1980).

11.2 DISPLACEMENT UNDER STATIC LOADS

11.2.1 General

Calculation of foundation displacements (usually axial settlements) should consider the following components:

- immediate displacements
- primary consolidation settlement
- secondary compression (creep) settlement
- settlements induced by cyclic loads
- regional settlements.

The first three components are the same as for shallow foundations (ISO 19901-4:2016, section 7.5.2.1). Differential settlements (induced by spatial soil variability, moments, torque and eccentricity) are not assessed for individual intermediate support foundations, as they are essentially rigid.

For routine design, soil type considerations include:

(a) Sands (and stiff clays and rock): settlement magnitudes are small and not problematic. Hence, settlements are not considered for routine design.
(b) Clays (fine-grained lithogenous sediments) are the most abundant of all marine sediments (about 70% by volume). These clays are usually normally consolidated (or at best lightly overconsolidated) and give the largest settlements.
(c) Organic soils (peats) are rarely encountered offshore.

Table 11.1 summarises the relative importance of each of these three settlement components for different soil types.
   Settlement considerations include:

(a) There is no experience with long-term settlements.
(b) The largest settlement component is judged to be consolidation; however, cyclic and/or creep components could be significant. For example, extreme wave loading on a stubby caisson in sand may degrade outer friction, causing untoward settlement of a non-underbase grouted support foundation. Similarly, OWT monopiles have strict tilt criteria for accumulated displacement due to omni-directional cyclic lateral load.
(c) Consolidation of a clay soil plug may be important; the plug is incompressible during immediate loading but undergoes subsequent primary consolidation.
(d) Settlement assessments should generally aim to maximise the largest component (usually soil plug compression). Hence, the most important soil parameters are magnitude of outer friction and soil compression indices.

Section 11.2.3 discusses primary consolidation models.

Table 11.1 Relative Importance of Settlement Components for Different Soil Types. Intermediate Support Foundations – Routine Design

Settlement Component	Soil Type			
	Sand	Clay	Organic Clay	(Weak) Rock
immediate	yes	possibly	possibly	yes
primary consolidation	no	yes	yes	no
secondary compression (creep)	no	possibly	yes	no
cyclic	possibly	possibly	possibly	Possibly for weak rock
regional	possibly	possibly	possibly	possibly

Note: Based on Table 5.1 of Holtz (1991)

11.2.2 Immediate displacement

References for methods to assess immediate, elastic displacements that account for non-uniform soil profiles (e.g. linearly increasing soil strength) and foundation embedment are provided in this section. The results are generally used to provide a range of foundation stiffnesses for use by structural engineers or to check the initial portion of "load-settlement" response from geotechnical finite difference/element analyses.

The complete 4 × 4 seafloor foundation stiffness matrix (Section 8.3 and Figure 8.6c) should be assessed. Note that the lateral stiffness sub-matrix should not be derived using either p-y curves for deep (pile) foundations or elastic solutions for flexible piles. Reasons include the fact that intermediate foundations are essentially rigid and generally larger in diameter. Instead, either elastic solutions or numerical analysis should be used (DNV OS-J101, 2014a).

In order to evaluate the terms appearing in the stiffness matrix, sources for useful elastic solutions (all continuum based) for statically loaded circular intermediate foundations are listed in Table 11.2). It is seen that only Doherty and Deeks (2003) provide all six stiffnesses – and in non-homogeneous soil – but only for embedment ratios L/D $\leq$ 2. Of these solutions, only Baguelin et al. (1977) include a weaker zone. Their equation 12 can be used to demonstrate that a disturbed soil annulus has only a limited effect on lateral stiffness k_{xx}. For example, a disk with diameter D = 5 m, outside boundary radius R = 15D, soil Young's Modulus E' = 100 MPa, and Poisson's ratio v' = 0.3 has an intact stiffness k_{xx} = 121 MN/m/m length. For a 0.5 m wide disturbed zone with E' = 40 MPa, k_{xx} reduces to 113 MN/m/m length, i.e. a small 7% decrease. This has implications for laterally loaded OWT monopiles installed by vibratory methods, where both the annulus width and soil modulus are likely to be greater than for driven foundations. That is, since lateral stiffness is largely due to the far field displacement regime, and most lateral displacements at depth remain in the elastic zone, driven and vibrated monopile lateral responses under HM load will not differ significantly.

Until Gupta and Basu (2016) and Arany et al. (2017), there were no elastic continuum solutions for laterally loaded OWT type rigid caissons (embedment ratio L/D $\leq$ 10) in non-homogeneous soil. Carter and Kulhawy (1992) were valid for L/D $\leq$ 10 but are only for uniform soil and Doherty et al. (2005) included non-homogeneous soil but for L/D $\leq$ 2.

There are no elastic solutions for deeply embedded rigid circular foundations under MH loading – this situation occurs at the base of a 1-D Winkler support model for OWT monopiles: Poulos and Davis (1991) give solutions for V load only and for MH load on an embedded vertical plate (their section 15.4), and Doherty and Deeks' (2003) caisson geometry (b) is valid only for L/D = 2.

Table 11.2 Elastic Continuum Solutions for Intermediate Rigid Foundations – Static Load

Shallow or Intermediate Foundation Type	Axial k_{zz}	Lateral k_{xx}	$k_{x\theta}$ $k_{\theta x}$	$k_{\theta\theta}$	Torsional $k_{\psi\psi}$	Reference	Remarks
Surface VHMT load	✓	✓	×	✓	✓/×	DNV (2014a)	Table G-1 Circular footing on stratum over bedrock or on stratum over half space
Embedded footing VHM load	✓	✓	×	✓	✓/×	DNV (2014a)	Table G-2 Circular footing embedded in stratum over bedrock L/D ≤ 1
Embedded footing arbitrarily shaped VHMT load	✓	✓	×	✓	✓	Gazetas (1991)	homogeneous half-space
Four geometries: (a) footing: at base of hole; (b) embedded footing; (c) solid caisson; (d) caisson VHMT load	✓	✓	✓	✓	✓	Doherty and Deeks (2003)	Rigid foundation Non-homogeneous half-space L/D ≤ 2
Solid caisson HM load	×	✓	✓	✓	×	Carter and Kulhawy (1992)	Corresponds to Doherty and Deeks (2003) geometry (c) in homogeneous half-space. L/D ≤ 10
Solid caisson HM load	×	✓	×	✓	×	Gupta and Basu (2016)	Multi-layered soil
Solid caisson HM load	×	✓	✓	✓	×	Arany et al. (2017)	Table 3
Pile V load	✓	×	×	×	×	Randolph and Wroth (1978)	Equation (30)
Disk H load	×	✓	×	×	×	Baguelin et al. (1977)	Equation (8)
Pile T load	×	×	×	×	✓	Randolph (1981)	Equation (9)

11.2.3 Primary consolidation settlement

According to Hernandez-Martinez et al. (2009), "No literature source has reported in situ long term settlement measurements for skirted foundations." Experience with monitoring gravity platforms whose foundation comprises skirted walls is available; Lunne et al. (1982) presented a comprehensive study showing settlement records obtained from skirted foundations of platforms in the North Sea for continuous monitoring periods that varies from approximately 7–8 months to 40 months. However, this experience could not be transferred directly to skirted foundation analyses because these platform measurements are restricted to small depth to width ratio. More recently, Svanø et al. (1997) reported settlement records from the Troll A platform foundation in the North Sea after 10 years of continuous monitoring.

Centrifuge tests provide an alternative to further investigate skirted foundation long-term settlement evaluation. So far, a number of centrifuge tests have been performed worldwide to better understand issues such as installation and capacity, but no emphasis has yet been placed on long-term settlement.

Chen and Randolph (2007a) reported results from a series of centrifuge tests where instrumented suction anchors (simulating a suction anchor prototype of 14.4 m length, 3.6 m diameter, 0.06 m wall thickness and 230 kN self-weight) were modelled in normally consolidated reconstituted kaolin. Although their main objective was to investigate the external radial stress changes around suction anchors only, they reported that after the suction anchor was installed to target penetration depth, followed by one hour of consolidation (equivalent to 1.7 years at prototype scale), the settlement of the suction anchor model was 0.10 mm (equivalent to 2 cm at prototype scale). This settlement was due, however, only to suction anchor self-weight. Larger settlement is anticipated once additional loads are considered as the case of manifold and pump stations.

The primary consolidation settlement model given in the following paragraphs is considered reasonable for routine intermediate support foundation design in clay: both data and model accuracy (Lambe, 1973) are similar.

Load transfer of seafloor load V is via outer skin friction, and the rest is taken by base load. The base load compresses the soil plug upwards and the soil below foundation tip level downwards. The corresponding foundation consolidation settlement model is shown in Figure 11.1 and consists of compressing two non-linear springs in series (Figure 11.1a). The top spring represents the soil plug and the bottom one the soil below tip level (embedment depth L).

Considering the foundation outer steel and soil plug base (Figure 11.1b), seafloor (permanent) load Q_{SLS} is resisted by mobilised outer skin friction F_o, and the remainder by base load ($Q_{base} = Q_{SLS} - F_o$), acting over the complete foundation area A_{base}. Figure 11.1c and 11.1d show the forces on the foundation steel and soil plug respectively for the usual case of no top plate bearing. Wall base load, Q_{wall}, is essentially zero (Figure 11.1c). Hence,

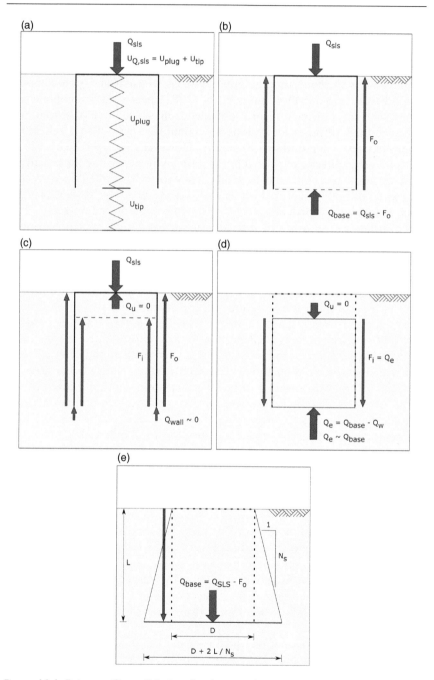

Figure 11.1 Primary Consolidation Settlement (support foundation): (a) simple two-spring model, (b) through (e) corresponding (SLS, permanent) loads and resistances; Soil Plug submerged weight not shown for clarity.

mobilised inner skin friction F_i, which acts upwards on the steel and downwards on the soil plug, has a value equal to Q_{base}. Considering loads on the clay soil plug (Figure 11.1d), the top spring is compressed upwards from tip level by a vertical effective stress increase $\Delta\sigma_v' = Q_{base}/A_{base}$. Due to inner friction resistance F_i, $\Delta\sigma_v'$ decreases with distance above tip level, and the corresponding u_{plug} (soil plug spring compression) final primary consolidation settlement value can be assessed using textbook equations (e.g. Holtz, 1991) for 1-D oedometer compression and soil parameters C_c, C_s (compression indices) and p'_c (preconsolidation pressure).

Additional settlement will occur due to increased vertical soil stresses in the vicinity of the foundation. Therefore, settlement assessment also requires calculation of the stress distribution in the soil below the foundation. Hence, u_{tip} (soil settlement below tip level) can be computed using loads at tip level and conventional load spread theory or similar. Distributions of $\Delta\sigma_v'$ (which may be assumed to be uniform) due to two loads which need to be superimposed. Figure 11.1e shows that they are (i) base load Q_{base} acting over a circle diameter D and (ii) mobilised outer skin friction F_o spread out with a load spread factor n_s to act over an annulus with inner diameter D and outer diameter $D + 2 L/n_s$.

For a fixed geometry, consolidation settlement magnitudes are sensitive to F_o, C_c, p'_c and C_s (in that order). If uncertainties exist, then analyses should consider a high estimate of u_{plug}. This is achieved by using a low estimate of F_o. The model presented here is useful for foundation geometries where plug final settlement may be significant, generally embedment ratios L/D less than around 3. Longer intermediate foundations behave more like pipe piles – tip settlement component u_{tip} dominates but is small at working loads.

For axially loaded support foundations in clay, settlement rates are a product of 1D vertical drainage within soil plug plus 2D drainage (combined vertical + radial outwards) below foundation tip level. The major unknown is the mass coefficient of consolidation value.

11.2.4 Secondary compression (creep) settlement

For fine-grained soils, in the absence of project-specific data, a reasonable value for secondary compression index C_α (slope of voids ratio e versus log time) is given by $C_\alpha/C_c \approx 0.04$, where C_c is the compression index (Terzaghi et al., 1996).

11.2.5 Settlement induced by cyclic loads

Janbu (1985) provides a (simple neat) cumulative strain model, while O'Riordan (1991) provide a useful review of the topic.

11.2.6 Regional settlement

Causes of regional settlement include (most common first):

- oil, gas and water extraction
- incomplete primary consolidation (caused by rapid sedimentation)
- ongoing tectonic movements.

11.3 DISPLACEMENT UNDER DYNAMIC AND CYCLIC LOADS

Do not use either pile cyclic p-y data (ISO 19901-4:2016a Section 8.5, soil reaction for piles under lateral loads) or elastic solutions for flexible piles for intermediate foundation design.

For OWT intermediate foundations (aka monopiles) the main geotechnical design challenge is that tilt should not exceed the allowable threshold. Typical SLS criterion is that accumulated tilt at seafloor < 0.5° A secondary challenge is eigen frequency assessment of complete OWT system (i.e. foundation, substructure, mast and rotor nacelle).

Monopile design requires a good knowledge of the soil structure interaction at small displacements. Offshore wind turbines are commonly designed so that their first natural frequency avoids frequencies of operational and environmental loads, see Figure 11.2 (CFMS, 2019). Design also needs to ensure that the permanent inclination of the foundation does not exceed a fixed limit at the turbine end of life (the inclination typically should remain < 0.5°).

From back analysis of monitored data, Kallehave et al. (2012) found current design models fail to reproduce the natural frequency although it is a key parameter for the design. Most soil-pile stiffness models derived for flexible piles tend to underestimate the natural frequency.

The method normally used to design piles under lateral and overturning loads is based on a Winkler modelling approach, commonly termed the p-y method. P-y curves proposed in standards API RP2A (API, 2000) or DNV-GL RP C212, (DNV-GL, 2017, 2019) are generally regarded as being satisfactory for jacket pile design but fail to provide reliable response for large diameter piles at small displacements. Several semi-empirical approaches have been proposed to stiffen the initial lateral response for "large diameter" piles, notably:

- Kallehave et al. (2012), Sorensen et al. (2010) and Kirsch et al. (2014) for sand
- Stevens and Audibert (1979) and Kirsch et al. (2014) for clay.

The initial response could also be calibrated on the basis of in situ measurements such as High Pressure Dilatometer (HPD) tests, or using finite

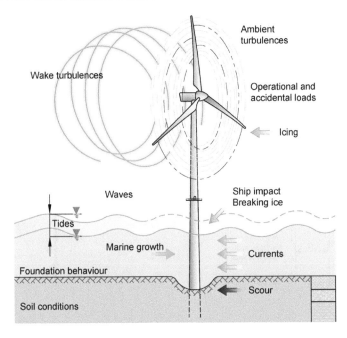

Figure 11.2 Schematic of environmental conditions to consider in modelling an offshore wind turbine (CFMS, 2019).

element analyses. DNV-GL RP C212 (DNV-GL, 2017, 2019) propose to use Stevens and Audibert (1979) to account for the "diameter effect" for piles up to approximately 2.5 m but highlight that alternative methods such as finite element (FE) analysis are needed for larger piles.

The PISA (Pile Soil Analysis) project (Byrne et al., 2020) proposed a new approach based on the calibration of soil reaction functions (1D model) using 3D FE analyses. The four soil reaction components included in the 1D model are (see also Figure 11.3):

- lateral resistance represented by p-y curves
- shear strength at the base of the pile called "base shear"
- axial shear stress along the shaft called "distributed moment"
- a rotational resistance called "base moment".

The contribution of the three last additional components start to be non-negligible when the L/D ratio decreases and becomes less than 3.

Within the framework of the PISA project, the four elements of the ground response have been quantified in dense sands (Burd et al., 2020) and stiff clays (Byrne et al., 2020). The current modelling procedure is limited to monotonic loading. Ongoing work is being carried out to extend the method to cyclic loading (unpublished as of this writing).

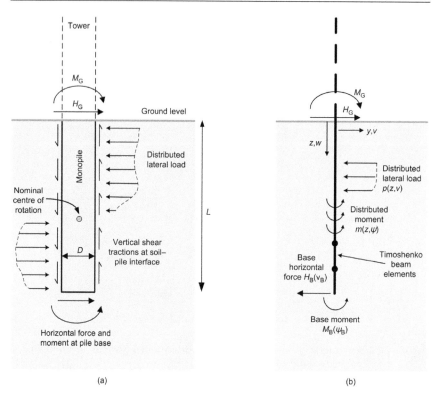

Figure 11.3 PISA model: (a) soil reaction components incorporated in the PISA design model; (b) 1D FE model employed in the PISA analysis.

Various methods have been proposed to derive the accumulated displacement (or rotation) under cyclic loading in sand (e.g. Lin and Liao, 1999; LeBlanc et al., 2010; Abadie et al., 2017). The SOLCYP JIP proposed a design approach for both sand and clay (SOLCYP 2017). Although, while initially focused on slender piles, the global approach accounts for the pile rigidity through a coefficient of rigidity, CR.

11.4 RESPONSE IN (WEAK) ROCK

Pile response in rock is most commonly determined through a local approach where the pile is modelled using a Timoshenko beam element (the Timoshenko beam model being more adequate for rigid piles than the Euler beam model) and the soil is modelled as a discrete set of independent springs along the length of the pile (i.e. springs operate in the horizontal plane, with no interaction with the springs above or below. The PISA project (Burd

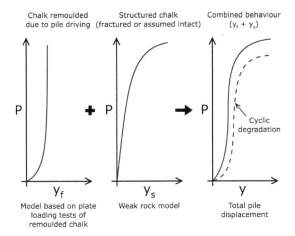

Figure 11.4 Illustration of chalk-pile load deflection concept (Muir Wood et al., 2015).

et al., 2020), highlighted that the so-called p-y method was originally developed for relatively slender piles and should not be sufficient to model the soil structure interaction of rigid piles with L/D < 6. However, in the case of weak rock, field experiences tend to show that properly calibrated p-y curves can predict reasonable pile rotation.

In the case of weak carbonate rock, pile installation (see Section 9.13) can lead to the creation of a crushed or remoulded zone around the pile. In that case, the subsequent lateral response will be softer than if the pile was surrounded by intact rock. To account for the presence of this remoulded zone, Muir Wood et al. (2015) proposed to consider two springs in series (Figure 11.4). The overall deflection of the pile (y) under a given load is assumed to be the sum of the deflection of the remoulded chalk (yr) and the deflection of the intact chalk (ys).

A similar concept was applied for driven and drilled and grouted piles in calcarenite (Lovera, 2019) where the grout annulus, crushed zone and intact rock were modelled as three springs in series (Figure 11.5).

The equivalent initial local stiffness can be deduced using

$$E_{equiv} = \frac{p}{y} = \frac{E_{oed}}{a \cdot \frac{\delta_{grout}}{\lambda \cdot D_{steel}} \cdot \frac{E_{oed}}{E_{grout}} + \frac{\delta_{crushed}}{\lambda \cdot D_{crushed}} + \frac{E_{oed}}{G_{rock}}} \quad (11.1)$$

where:

For the grout,

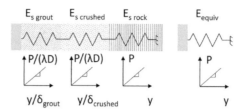

Figure 11.5 Model for pile response in weak rock with grout anulus, crushed zone and intact rock modelled as three springs in series (Lovera 2019).

a parameter equals zero for a driven pile and 1 for drilled and grouted piles;

D_{steel} is steel tubular diameter;

δ_{grout} is grout annulus thickness;

E_{grout} represents shear modulus of the grout;

λ parameter is smaller than one, taking into account that the pressure does not apply on the all circumference.

For the crushed/remoulded zone,

$D_{crushed} = D_{steel}$ for driven pile and D_{grout} drilled hole diameter for drilled and grouted piles;

$\delta_{crushed}$ is the crushed zone thickness;

E_{oed} is the oedometer modulus of the crushed zone material at low stresses.

For the rock,

G_{rock} is the shear modulus of the rock.

Cyclic degradation under lateral loading is generally marginal for non-carbonate competent rocks. For weak carbonate rocks, experience shows that significant accumulated displacements can be observed. These accumulated displacements are due to both the compaction of the crushed zone and the cyclic degradation of the rock mass. Based on observations made during field tests (Palix and Lovera, 2020), Lovera et al. (2020) proposed a rheological model to predict the accumulation of cyclic displacements based on an analogy with the creep phenomenon, illustrated in Figure 11.6.

At the element scale, deformations under cyclic loading and creep are shown to share some similarities. Under both creep and cyclic loading, the sample is weakened and can fail at a lower level of loading than under

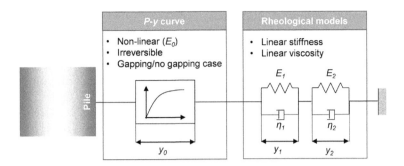

Figure 11.6 Sketch of the rheological model considered for piles driven in soft rock (Lovera et al., 2020).

monotonic conditions. For a high maintained load (similarly for a high maximum cyclic load), the strain accumulates faster and faster, leading quickly to failure. For a low maintained load (similarly for a low maximum cyclic load), the strain accumulates at the beginning and then stabilises.

The Lovera et al. (2020) model combines the non-linear static p-y curves with Kelvin-Voigt models in series. A Kelvin-Voigt model involves two parameters: a stiffness (E1) that controls the maximum accumulated displacement when the stabilised state is reached and a viscosity η1 that controls the rate at which the stabilised state is reached (the higher η1, the slower the stabilised state is reached). For driven piles, two springs are considered: one for the crushed rock for which the accumulation of displacement (y1) stabilises quickly (low value of η1) and one for the weak rock for which the accumulation of displacement (y2) is more progressive (high value of η2).

11.5 CLOSURE

This chapter has discussed in-place response and has shown that different intermediate foundation types have their own challenges. Anchor foundation movements are usually non-critical and are rarely assessed.

Support foundations may be conveniently subdivided into those for oil and gas and renewable projects. The challenges are listed below in frequency order (most commonly encountered first).

Support platform projects require 4 × 4 foundation stiffness matrices to pass to the structural engineer. Simple (elastic) solutions have been listed to evaluate the stiffness terms k_{zz} (axial), k_{xx}, $k_{x\theta}$, $k_{\theta x}$, $k_{\theta\theta}$ (lateral, coupled) and $k_{\psi\psi}$ (torsional). ISO/API type "p-y" curves are unsuitable for intermediate foundation lateral response assessments due to the rigid body rotational behaviour of intermediate foundations.

Renewable support foundations, especially OWT monopiles, require analysis of both tilt and Eigen frequencies.

Support platform projects with high mobilised axial V/V_{max} values in weak clay profiles may require analysis, particularly of primary consolidation settlement magnitude. To this end, a reasonable two-spring model has been presented.

The Doherty and Deeks (2003) continuum solutions are comprehensive, covering a wide range of geometries and non-homogeneous elastic half-space profiles routinely encountered in design. However, they are valid only for embedment ratios $L/D \leq 2$. It would be nice if their work could be extended to cover higher embedment ratios as OWT monopiles have L/D values up to 10.

Chapter 12

Miscellaneous design considerations

12.1 INTRODUCTION

An overview of the miscellaneous design considerations for intermediate foundations discussed in this final chapter are shown in Figure 12.1. Some are non-geotechnical, reflecting the necessary interaction between other off-shore engineering disciplines (structural, installation, hydraulic, etc.).

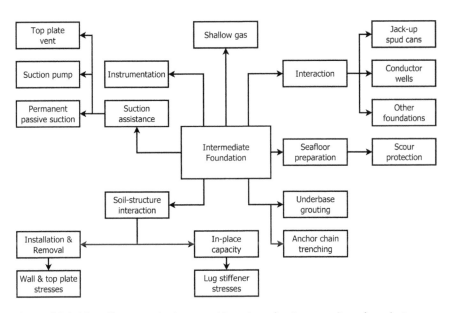

Figure 12.1 Miscellaneous design considerations for intermediate foundations.

12.2 SCOUR PROTECTION

The in-place resistance of intermediate foundations with low embedment ratios (less than around 2) in sand is generally sensitive to seafloor scour. If scour is a potential problem, then scour protection is generally recommended as an alternative to regular seafloor/foundation inspections. Protection types include frond mats, rock fill, sandbags, or fences and diversion berms.

If there is a significant time lag between foundation and scour protection installation, additional in-place analyses are normally performed of the unprotected foundation to verify that factored design loads do not exceed the factored soil resistance. Such analyses use a specified scour geometry, for which soil strength values outside and below the foundation may be reduced, plus environmental VHMT loads and load factors γ_L appropriate to the time period considered.

In lieu of project-specific data, the general and local scour default recommendations given for pipe pile foundations in sand (ISO 19901-4:2016a, Section A.8.5.2), namely local scour depth equal to 1.5 D and 6 D overburden reduction depth, are inappropriate for intermediate foundations. Whitehouse et al. (2011) review scour development around monopiles in sand for nine offshore sites and show that scour protection slightly decreases scour depth and puts it further away from the pile (compare their Figures 2a and 2b). DNV (2014a) gives general comments on scour around monopiles and equations to calculate local scour in their Appendix J. At the time of writing there are no publications on scour around large diameter foundations (e.g. D > 7.5 m).

12.3 INTERACTION WITH ADJACENT INFRASTRUCTURE

12.3.1 Introduction

Intermediate foundations may interact with the following adjacent infrastructure (the most important are listed first):

- jack-up spudcans
- adjacent intermediate foundations
- adjacent shallow foundations and pipelines
- conductor wells.

The remainder of this section gives information about interaction between intermediate support foundations and other adjacent infrastructure, located either on or beneath the seafloor. At the time of writing, only one publication in the public domain covers interaction of intermediate foundations with jack-up spudcans (Alderlieste and Dekker, 2018). Hence, most information is extrapolated from other studies.

12.3.2 Jack-up spudcans

Minimal-facilities hydrocarbons platforms normally require subsequent location of a jack-up rig to install well conductors. Similarly, offshore wind turbine monopiles or tripod substructures may be installed using a jack-up rig fitted with spudcans. Figure 12.2 shows examples. Hence there may be a risk of overstressing platform foundations. Such situations require consideration of interaction between the jack-up spudcan(s) and platform foundations, particularly in weak soil profiles. Geotechnical analysis involves assessment of soil movements induced by spudcans and the effects of these movements on

(a)

(b)

Figure 12.2 Examples of jack-up spudcan – Intermediate foundation interaction: (a) Work-over jack-up and Ophir well head platform (Alderlieste and Dekker, 2018) and (b): installation jack-up vessel (Courtesy Fred. Olsen Windcarrier).

the foundation, both lateral and axial (Mirza et al., 1988; Siciliano et al., 1990). Normal practice for pipe pile foundations in normally consolidated clay profiles is to ignore interaction effects provided that the closest soil gap distance is at least one spudcan diameter (ISO 19905-1:2016b).

However, for intermediate foundations, this gap distance (1 spudcan diameter D) should be critically reviewed. This is because the foundations are essentially rigid: any load due to lateral soil displacement is transferred directly into the platform leg, and not partially absorbed by pile bending.

A key issue is predicting lateral soil movements induced by the spudcan at the intermediate foundation. These movements have both horizontal and rotational components. Due to smaller foundation embedment depths (than piles), rotational effects are usually more critical than lateral effects – foundation rotations could induce additional BM/overstress the platform leg. This situation is similar to kicking one leg of a three-legged stool.

Jack-up spudcan–foundation interaction analysis is difficult. This is because the accuracy of any prediction is a function of the accuracy of both the data and the model (Lambe, 1973), and larger-than-usual inaccuracies generally exist both for data (e.g. soil shear strength profile, spudcan geometry and preload, foundation geometry and gap distance) and the model (i.e. spudcan penetration tip depth and resulting lateral soil movements around the spudcan).

Spudcan installation induces vertical foundation movements. However, industry perception (for pile foundations) is that axial (downdrag/upthrust) is not so critical as lateral. The majority of axial pile resistance is derived from soil well below the spudcan and is rarely (possibly never) studied in detail.

Hence, instead of complex 3D large strain Finite Element Analyses, a reasonable design approach may be to:

- assign cautious parameter values for data
- assess Low, Best and High Estimate foundation "p-y" data, either from analytical equations (Section.11.2) or from 3D Finite Element Analyses
- assess cautious (high) spudcan penetration and corresponding lateral soil movement profile at intermediate foundation centreline
- offset foundation p-y data using lateral soil movement profile
- use offset intermediate foundation stiffness matrices in platform structural analyses.

Note that care has to be taken with orientating foundation p-y data. This is because the offsets need to be applied along a line connecting the spudcan and the jacket leg(s), whereas the platform structural model automatically rotates these to coincide with the storm/wind direction. However, since spudcan penetration is unlikely to occur during a "small" (10-year summer storm) event, then orientation may not be critical.

Further information on this approach is given by Xie (2009) and Xie et al. (2017) on lateral soil movement data, by Tho et al. (2013 and 2015) on decoupled y-shift method, by Dekker (2014) on foundation stiffness matrix using two lateral springs and by DNV-OS-J101 (DNV, 2014b), which deprecates pipe pile p-y curves and suggests using FEA instead.

Note that, unlike piles, additional caisson wall thickness increases are unnecessary – the caisson is almost rigid (and hence capable of withstanding high BM) and lateral soil pressure increases are small (e.g. < 10 kPa). Nevertheless, one should verify that foundations can withstand additional soil pressures.

Like spudcan-pile interaction, it is likely that spudcan-intermediate foundation interaction in competent soils is not usually a major concern. However, in weak soil, where deep spudcan penetration occurs (≥ 0.5 D, say), then detailed jack-up spudcan – foundation interaction analysis may lead to considering:

- increased platform leg stiffening/braces
- increased foundation wall thickness
- minimizing spudcan diameter and penetration
- maximizing spudcan – foundation clearance
- using a mat supported (instead of an independent leg) MODU (mobile offshore drilling unit)
- using the jack-up rig to install the platform (i.e. install foundations into pre-shifted soil)
- pile (instead of intermediate) foundations.

Another issue is possible long-term intermediate foundation lateral resistance decrease due to the presence of remoulded soil/footprint. This was studied by Stewart (2005) for spudcan–pile (not intermediate) foundation interaction.

Alderlieste and Dekker (2018) have described detailed jack-up–suction pile interaction design studies for the Ophir Well Head Platform (WHP), offshore Malaysia. Due to the weak soil profile (Figure 12.3), and the high spudcan preload, penetration depths between 7 m and 15 m were possible. For the design scenario (Figure 12.4), these resulted in "free-field" soil lateral displacements of 0.12 m and 0.19 m at seafloor and suction pile tip level respectively. Even though the soil gap spacing (1.1 D) exceeded the ISO 19905-1:2016 recommendation, these displacements were sufficiently large to necessitate detailed studies. Instead of using shifted p-y curves, a more rational (less complicated) approach was used to obtain the suction pile loads and displacements. Figure 12.5 shows WHP displacements: after installing the starboard spudcan, only the closest suction pile undergoes "free-field" soil lateral displacements (the other two suction piles are too far away), and jacket rotation in plan occurs. As expected, symmetrical pile displacements occur after installing the second (port) spudcan. The back suction pile furthest away is the most highly loaded, since it restrains jacket movements. No additional displacements are induced by the bow spudcan. Based on their results, they concluded that 1 D minimum soil gap spacing is adequate.

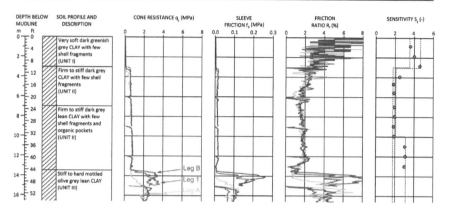

Figure 12.3 Ophir well head platform, soil properties (Alderlieste and Dekker, 2018).

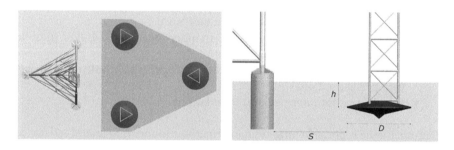

Figure 12.4 Top view (left) and side view (right) of work-over jack-up rig at Ophir well head platform. Suction pile diameter D = 6 m, target penetration L = 13.7 m. Spudcan diameter D = 17 m, spudcan design penetration h = 12.75 m, minimum spudcan-suction pile clear soil gap S = 19.4 m (Alderlieste and Dekker, 2018).

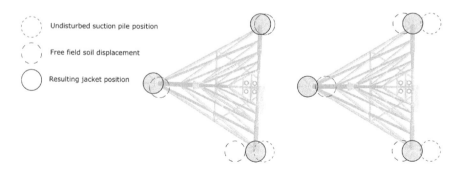

Figure 12.5 Suction pile displacements due to spudcan penetration (not to scale) at Ophir well head platform. Left: first spudcan installation. Right: due to second spudcan installation (Alderlieste and Dekker, 2018).

12.3.3 Adjacent intermediate foundations

In-place resistance and response both should consider the possibility of interaction with any adjacent intermediate foundations.

Considering block failure in clay, design failure zones should not intersect. Hence, a general rule of thumb for undrained soil response using 45 degree wedges is a clearance of at least 1.5 times the foundation embedded length. Because of curvature of the passive failure surface, this clearance value should be increased for sands.

Finite element analysis is advisable to capture the effect of complexities in stratigraphy or geometry.

12.3.4 Adjacent shallow foundations and pipelines

In-place resistance and response of intermediate foundations should both consider the possibility of interaction with any adjacent shallow foundations and pipelines.

Interactions are discussed in Engin et al. (2015) for jack-up–mudmat (seafloor template), Versteele et al. (2015) for mat-supported MODU–pipeline, and Kellezi et al. (2005) for spudcan-pipeline. Since an intermediate foundation installation displaces less soil (in the order $10 m^3$) than a jack-up spudcan (in the order $150 m^3$), interaction effects on a pipeline are expected to be less from an intermediate foundation than from a spudcan (at the same soil gap spacing).

12.3.5 Conductor wells

Conductor installation techniques should consider the possibility of soil removal (wash-out) occurring in sand and appropriate measures (e.g. using driven isolation casings) taken. This is to prevent loss of resistance at nearby intermediate foundations.

Intermediate foundations also interact with the conductor wells. Generally, the only potential problem is that, when laterally loaded, intermediate foundations may induce additional displacements in the conductors. However, due to their flexibility, they will probably accommodate the extra lateral soil movement without significant bending moment increases.

12.4 SHALLOW GAS

The possibility of shallow/dissolved gas adversely affecting in-place resistance should be considered.

Geotechnical design of intermediate suction foundations where dissolved gas is present in clay includes the following considerations:

- Site investigation: gassy soils are difficult to sample and test in the laboratory. Therefore, site investigations should include in-situ tests.

- Installation: since such soils are "softer", and the vertical stress reduction may cause gas to come out of solution, more heave than usual may be anticipated. However, this may be offset by the fact that installation underpressures (i.e. tensile forces on the soil plug) are likely to be low in normally consolidated soft clays.
- In-place response: in time, gas migration could create a gas "bell" under the foundation top plate, causing an upwards force. This potential problem is removed by providing open vents in the top plate. Assuming vents are present, foundation compressive resistance can still accommodate bearing pressures under the foundation top plate. Maximum axial tensile pull-out resistance, V_{max}, cannot rely upon "passive suction" (since the top plate vent is permanently open). Instead, pull-out should be based on outer friction plus the minimum of (inner friction or soil plug weight) plus foundation submerged weight only, i.e. the "coring" and "leaking" failure models shown in Figure 10.1.

12.5 PERMANENT PASSIVE SUCTION

The use of permanent underpressure may be considered in order to increase foundation axial tensile resistance. This is particularly for sand profiles and support foundation platforms (underpressure can be applied by reducing water level within legs). Permanent underpressure ("passive suction") removes some of the uncertainties associated with temporary passive suction generated under the top plate discussed in Section 10.5.3 (on undrained "sand" soil response). The top plate vent should be open to allow the pump to remove water entering the soil plug. Pump design and seafloor outflow rate are discussed in Section 12.7. The downside is that underpressure has to be available, either permanently or during storm loading, throughout the lifetime of the structure.

12.5.1 Example – permanent passive suction

An example assessment of the additional compressive force (ΔV) and flow rate Q (assuming maximum underpressure) has been made for the following situation: caisson D = 10 m, L = 7.5 m, high estimate of sand permeability k = 2e-4 m/s, with a low estimate of submerged unit weight, γ_{sub}= 9 kN/m³. From Figure 9.16 (on comparison critical suction pressures), a low estimate of $\Delta u_{crit}/(\gamma_{sub}D)$ value is 1.0, giving Δu_{crit} = 90 kPa, corresponding to ΔV = $(\pi D^2/4)* \Delta u_{crit}$ = 7 MN. The flow rate Q can be found using Figure 9.5b and the equation $Q = (k D \Delta u/\gamma_{sub})$. Dimensionless factor F is a function of (L/D) and k_r, the soil plug: soil mass permeability ratio. A high estimate F value is 2.5, giving a (high estimate) Q value of 45 litres/s. For comparison, filling a bath requires approximately 0.6 litres/s, and a 100% redundant suction pump system can provide around 80 litres/s. Both ΔV and Q are considered reasonable. Better (and hopefully lower) Q estimates may be possible by careful examination of suction pump instrumentation data (penetration rate and pump outflow rate) during installation, especially at/near final penetration, where steady state conditions are most likely to have been achieved, and the water plug volume decrease is small: subtracting the water plug volume decrease from the pump outflow gives Q, the amount of water entering the soil plug.

Laboratory model tests on "vacuum anchors" have been reported by several researchers during the 1980s, including Goodman et al. (1961), Brown and Nacci (1971), Wang et al. (1975, 1977, 1978) and Helfrich et al. (1976). Wang et al. (1977) give an equation for (vertical) breakout capacity. Permanent underpressure is currently not a relied-on solution.

12.6 TOP PLATE VENT DESIGN

Vent(s) in the top plate are required to expel the "water plug" (the volume of water trapped between the top plate and seafloor) during the self-weight penetration phase. The vent outlet pipes should be of sufficient diameter to prevent high overpressures occurring in the "water plug" for the specified range of penetration velocities. If water pressures are high, they may cause piping along the foundation perimeter (sand profile) or base failure (clay soil profile). If the foundation design relies on in-place "passive suction", or extraction using overpressure is envisaged, then vent design should incorporate a valve. The valve has to be open during installation, and subsequently closed. This is to ensure a permanent seal.

For intermediate foundations installed by crane, an additional vent design consideration may be while the foundation is being lowered through the wave splash zone: vents should also be sufficiently large to allow essentially free air egress/ingress and hence avoid "snatching" (large variations in lifting wire loads).

The basic equation for incompressible fluid flow through a single vent (nozzle) is given by the equation:

$$v_n = C_d \sqrt{(2\,g\,\Delta H)} \qquad (12.1)$$

where

v_n = vent (nozzle) velocity   [m/s]

C_d = vent (nozzle) discharge coefficient   [–]

g = acceleration due to gravity   [m/s^2]

ΔH = vent (nozzle) differential head loss   [m]

Hence, for a single vent, the overpressure increase, Δu, and corresponding vent (nozzle) resistance R_n, are given by:

$$\Delta u = \Delta H\, \gamma_{water} \qquad [kPa] \qquad (12.2)$$

$$R_n = \Delta u\left(A_i - A_n\right)[kN] \qquad (12.3)$$

where

ΔH = vent (nozzle) differential head loss [m]

$\quad\;\; = [1/(2\ g)]\ (v_n^2/C_d^2)$

γ_{water} = fluid (sea water) density (= 10.05) [kN/m³]

A_i = pile internal area $(\pi\ D_i^2/4)$ [m²]

A_n = vent (nozzle) area [m²]

g = acceleration due to gravity [m/s²]

v_n = vent (nozzle) velocity [m/s]

$\quad\; = v_{pile}\ A_i/A_n$

v_{pile} = foundation velocity [m/s]

C_d = vent (nozzle) discharge coefficient [–]

Equation (12.2) can also be used to assess the vent (nozzle) area A_n required to prevent the overpressure Δu causing base end-bearing ("push-out") failure. In addition to the foundation velocity V_{pile}, top plate vent area A_n results are sensitive to the coefficient of discharge, C_d. The minimum C_d value should be taken to maximise vent (nozzle) resistance R_n. If the vent is sharp-edged and circular, the "ideal" C_d value is 0.611 (= $\pi/(\pi + 2)$). Hence a cautious value is usually taken to be 0.61. Since entry into the seafloor is usually the most critical stage, then it is usually sufficient to analyse this situation only. For controlled lowering, a reasonable value for foundation touch-down velocity, v_{pile}, is 0.2 m/s (see Section 9.9).

12.6.1 Example – top plate vent design

As an example of the use of Equations (12.2) and (12.3), verify that the vent design (2 number 1.0 m diameter, each with area 0.8 m²) will not cause end-bearing failure at seafloor. The relevant data, which are similar to those used in the Section 9.9.3 (free-fall penetration) Case A example, are as follows:

$D = 5.0$ m, $WT = 25$ mm ($A_i = 19.2$ m²)

$N_{c,circle} = 6$

$s_{u,0} = 2$ kPa

$FOS = 1.5$

seafloor velocity v_0 = 0.2 m/s

two vents, diameter = 1.0 m

A_n = 0.8 m^2

$n_{nozzles}$ = 2

vent (nozzle) discharge coefficient C_d = 0.61

Actual and allowable overpressures Δu and Δu_{all} are evaluated as follows:

vent (nozzle) velocity v_n = 2.41 m/s (= $v_{pile} A_i/(A_v n_{nozzles})$)

vent differential head loss ΔH = 0.79 m (= $1/(2 g)][v_n^2/C_d^2]$)

actual overpressure Δu = 7.8 kPa (= $\Delta H \gamma_{water}$)

maximum allowable overpressure Δu_{all} = 8 kPa (= $N_{c,circle} s_{u,0}$/FOS).

Based on these results, the vent design is satisfactory ($\Delta u < \Delta u_{all}$).

As stated earlier, the seafloor situation is assumed to be the most oner-ous. More complicated calculations are necessary should velocity increase with depth below seafloor, for example due to free-fall caused by uncon-trolled lowering (Section 9.9.3, free-fall penetration, and Figure 9.8, free fall velocity – depth curve). In such cases, the R_n, the corresponding vent (nozzle) resistance, has to be computed, and interested engineers may care to check that R_n is 143 kN when v_{pile} = 0.2 m/s.

12.7 SUCTION PUMP DESIGN

During installation, the suction pump maximum achievable flow rate should be sufficient to discharge both the "water plug" volume plus the anticipated outflow rate at seafloor. This is especially important in sands and gravels of high permeability and large diameter foundations (e.g. > 10 m).

Furthermore, the suction pump used during installation should be capa-ble of generating adequate underpressure/overpressure.

To calculate the "water plug" flow rate component, a penetration velocity of 1 mm/s is generally used. This is a factor 20 slower than a CPT. Houlsby and Byrne (2005) give an equation for assessing seafloor outflow rate due to underpressure. A high estimate of soil permeability should be used.

12.7.1 Example – suction pump flow rate

A flow rate Q assessment is made for the example previously studied in Section 12.5 (permanent passive suction). Relevant Q data are: caisson D = 10 m, WT = D/200 = 50 mm, L = 7.5 m, sand permeability k = 2e-4 m/s, with a high estimate of seafloor outflow $Q_{seafloor}$ = 45 litres/s.

Inner diameter $D_i = D - 2\,WT = 9.9$ m

Water plug area $A_i = \pi\,D_i^2/4 = 77.0$ m^2

Assume penetration velocity $v = 0.001$ m/s

Pump flow rate to discharge water plug $Q_{plug} = v\,A_i = 0.077$ m^3/s $= 77$ litres/s

Pump flow rate to discharge seafloor flow $Q_{seafloor} = 45$ litres/s

Required minimum $Q = 77$ litres/s $+ 45$ litres/s $= 122$ litres/s (440 m^3/hour)

12.8 FOUNDATION INSTRUMENTATION

Provided that they are of sufficiently high quality, back-analyses of suction foundation installation records assist in model refinement/calibration/ extension. Examples include Colliard and Wallerand (2008) and Chatzivasileiou (2014). In general, particularly for installations in sand, underpressure data should be reliable: there should be no concerns about possible friction losses in the exhaust system between the top plate and the point where pressures were measured. In addition, if the preload value is large, then it has to be accurately known. In this respect, a difficulty for braced jackets (where all foundations are installed at the same time) is to assess the preload taken by each foundation: usually only crane hook load data are available.

Installation foundation instrumentation usually comprises measurements of:

- penetration
- ambient water pressures acting externally on the top plate and internally (either under the top plate or at the suction outlet pipe, rather than at the pump, in order to avoid errors due to pipe friction and Venturi effects)
- soil plug heave
- inclination and north-seeking gyroscope (intermediate anchor foundations)
- position (x, y, z co-ordinates)
- preload.

Preload measurements should be made both in air and under water using a calibrated load cell mounted between the crane wire and the hook. The foundation steel surface condition (both inside and outside) should also be documented.

Long-term foundation instrumentation is a growing area for offshore wind farms, where extensive monitoring systems, both short and long term, are being used to optimize foundation design (e.g. Devriendt et al., 2013; Byrne et al., 2015a, 2015b; Shonberg et al., 2017). In general, instrumentation should also be considered to improve geotechnical knowledge about foundation response during storm loading (Svanø et al., 1997).

12.9 STEEL DESIGN

Structural steel designers should also check the foundation geometry. Critical items normally include stress concentrations and fatigue in the vicinity of the load application point, and buckling of the steel cylinder and top plate under applied underpressure/overpressure.

In addition, foundation steel tip integrity assessments (e.g. Aldridge et al., 2005; Erbrich et al., 2010) should be considered if there is a risk of the foundation tip locally encountering an isolated gravel (or cobble or boulder-sized) fragment (see Section 9.10.5), and this fragment cannot be fractured (i.e. split longitudinally) or pushed to one side (i.e. bearing capacity failure of the soil surrounding the fragment).

If required, like soil reaction stresses (Section 12.10), steel stresses may be assessed by 3D non-linear FEA. However, the steel response is non-linear, whereas it is generally sufficient to assume that the soil remains elastic. Usually structural analysis software (SACS, Nastran, SESAM, ANSYS, etc.) is used instead of geotechnical analysis software (e.g. PLAXIS or OptumG2/G3) to determine steel stresses. Soil continuum elements (not a bed of Winkler springs) should be placed both inside and outside the foundation. If the complete intermediate foundation is being modelled, then a finite element mesh extending 3L radially and 2L vertically should be sufficiently accurate (Kuhlemeyer, 1979). If infinite elements are used, then the *finite* mesh radial extent can be reduced to 2L (Kay and Palix, 2010). Table 12.1 provides reasonable soil E and ν values.

For model verification, mesh nodes have 6 DOF (loads F_x, F_y, F_z, M_x, M_y and M_z and displacements δ_x, δ_y, δ_z, φ_x, θ_y and Ψ_z), Figure 3.1a gives their sign convention, and the RP is at seafloor on the foundation centreline. Assuming lateral loading is dominant, a simple way to verify the model is to run two load cases. The first applies a "fixed head" unit lateral displacement ($\delta_x = 1$ mm, remaining 5 DOF = 0) to the RP to find pile head lateral seafloor stiffness matrix terms k_{xx} (= F_x/δ_x) and $k_{x\theta}$ (= M_y/δ_x). The second load case finds the missing two terms $k_{\theta x}$ (= F_x/θ_y) and $k_{\theta\theta}$ (= M_y/θ_y) by applying "fixed head" unit rotation ($\theta_y = 1$ milliradian, remaining 5 DOF = 0). Off-diagonal terms $k_{x\theta}$ and $k_{\theta x}$ should be essentially identical. All four lateral stiffness matrix terms can then be compared with solutions in Table 11.2 (Elastic Continuum Solutions – Static Load).

12.10 SOIL REACTIONS

Soil reaction stresses are needed by structural steel designers as input to structural design of suction foundations. Usually, the following three phases need to be considered:

- installation and retrieval
- operation (in-place)
- removal.

Table 12.1 Soil Parameter Values for 3D FEA of Foundation Steel

Soil Type	Typical Values	Reference(s)
Clay	$E = 200\ s_u$ $\nu = 0.49$	Poulos and Randolph (1983)[a]
Sand	$E = 4\ q_c$ (NC) and $10\ q_c$ (OC) $\nu = 0.25$	Lunne et al. (1997)
Rock	$E = 350\ \sigma_c$ $\nu = 0.1$	Attewell and Farmer (1976)

Notation:
E = Young's Modulus
ν = Poisson's Ratio
s_u = soil undrained shear strength
q_c = CPT cone tip resistance
σ_c = unconfined compressive strength

Note:
[a] For laterally loaded piles.

However, during installation, retrieval and removal phases, general experience is that suction/excess water pressures govern steel stresses, and hence soil reaction stresses are not critical. In such cases, they are derived only for the operational phase.

If required, soil reaction stresses are normally assessed by 3D non-linear FEA. This technique has largely replaced the use of simplified analytical solutions, considering separate stress distributions due to different load components. The separate components are then combined to give design soil reaction stresses. Examples are given by NGI (1997).

12.11 UNDERBASE GROUTING

At the end of installation, a water-filled void may exist beneath the top plate. This void generally occurs in sand profiles. The presence of a void is generally due to using a high estimate stickup value, primarily to ensure that the target depth could be successfully reached. It is usually not practical to embed the foundation beyond its target depth. Reasons include (a) the void may be non-uniform and the (sand) plug may be in contact with the top plate underside and/or internal stiffeners. In such cases, additional penetration may induce high underpressures and steel stresses, and (b) the jacket will be (slightly) lower than intended, which may cause issues with pipelines and other connections.

Because of the limited foundation embedment ratio (usually less than one), there is a perceived risk of additional axial settlement occurring during cyclic storm loads. Hence, underbase grouting is generally applied. The objective is to replace the water in the void with cement grout.

A consequence is that such foundations cannot be removed by the application of overpressure (unless a hole is drilled through the cement grout into the sand plug prior to applying overpressure).

After curing, the cement grout should ideally be of similar stiffness and strength to the in-situ seafloor soil. If compartments (or inverted T-beam stiffeners under the top plate) exist, and there is only single grout entry and exit vents, then "rat-holes" or similar must be provided in order to minimise the risk of not completely filling the water-filled void. These considerations are similar to those for large gravity base foundations (i.e. shallow foundations) given in ISO 19903 (ISO 2019).

12.12 ANCHOR CHAIN TRENCHING

Anchor chain trenching of taut and semi-taut mooring systems is a relatively recent observation. Bhattacharjee et al. (2014) described wire rope and chain chafing integrity issues for the Sepentina FPSO taut anchoring mooring system, offshore Equatorial Guinea in 475 m water depth. More importantly, anchor chain trenches of varying dimensions were subsequently revealed in front of all nine suction piles; see Figure 6.6. Based on soil loss, the piles were considered unfit for service. The replacement was a (temporary) catenary mooring system comprising OMNI-Max free-fall anchors.

Bhattacharjee et al. (2014) provided the first mention in the public domain of anchor chain trenching. It is understood that similar trenches, some of which reached lug level, have also been observed in weak NC clays in other parts of the world, including the Gulf of Mexico and South-East Asia, for taut and semi-taut mooring systems.

Anchor chain trenching 3D FEA by Hernandez-Martinez et al. (2015) considered a typical Gulf of Guinea (GoG) soil profile, namely $s_{u,C}$ = 2.4 + 1.57 z [kPa, m] and γ_{sub} = 3.5 kN/m^3. The anchor geometry was D = 5 m, L = 20 m, with z_{lug} at 13.3 m bsf (i.e. 2/3 L) on the anchor outer surface. In plan, this was simplified to an equivalent rectangular area (3.92 m by 5 m) having the same cross-sectional plan area (i.e. 19.6 m^2) as the actual anchor. The chain trench geometry was "brick"-shaped – 5 m (1 D) wide down to lug level in front of the anchor and with a flat trench base. For typical taut anchoring systems, θ_{lug} typically varies between 40° and 20°, for which they obtained anchor holding capacity reductions of 20% and 45% respectively, as shown in Figure 12.6.

Arslan et al. (2015) also considered a rectangular trench with the same width as the foundation, but in front of a 5 m diameter circular foundation, and two embedded lengths, L = 14 m and 16 m. The generic soil profile was $s_{u,z}$ = 2 + 1.5 z [kPa, m]. The corresponding soil γ_{sub} value was not reported. Their limited parametric study indicated that similar capacity reductions (20% to 40%) are likely.

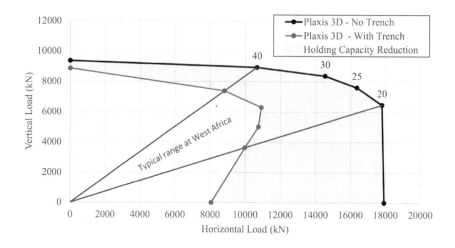

Figure 12.6 Anchor chain trench 1D wide. V_{lug}-H_{lug} envelopes (Hernandez-Martinez et al., 2015).

Alderlieste et al. (2016) adopted an improved approach, using 3D FEA of a 2 m wide sloping trench in front of a circular 6 m diameter, 19.35 m embedded length suction pile, for detailed design of a taut mooring anchor system in Gulf of Guinea clay. Figure 12.7 summarizes their V_{lug}-H_{lug} capacity envelopes, together with factored ALS loads. It is seen that there is a 4% reduction for ultimate vertical capacity, and 20% reduction for ultimate horizontal capacity. The latter reduction is significant, but decreases to approximately 7% for the design load angles, which were at least 30°.

Figure 12.8 shows PLAXIS 3D incremental shear strain results for the "fixed head" condition, and the corresponding interpreted failure mechanism. The main feature is a cylindrical rotational failure mechanism, resembling a pair of curved, partly closed circular elevator doors, extending from seafloor down to the sloping trench base. Below this, there is a passive wedge failure of limited extent, starting at pile tip level and exiting on the sloping trench base.

Sassi et al. (2017) conducted 1 g and centrifuge tests of chain-trench-soil interaction in both kaolin and remoulded Gulf of Guinea clay in order to provide a conceptual framework for anchor chain trenching. In a companion paper, Versteele et al. (2017) give details of proprietary CASCI software, an extended 2D chain-soil model with erosion, trench and tunnel stability mechanisms.

New field evidence was presented in a pair of OTC papers by Colliat et al. (2018) and Sassi et al. (2018). Both papers are Gulf of Guinea soil and met-ocean specific. Colliat et al. (2018) present and discuss anchor trench bathymetry data (width, depth and anchor-to-trench distances) for three

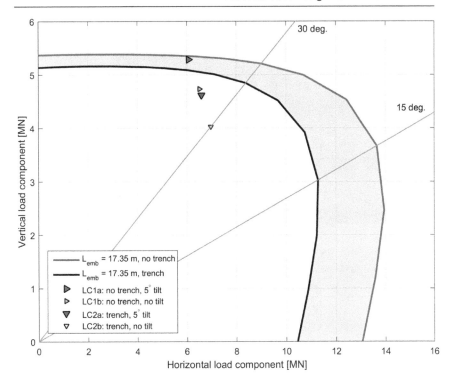

Figure 12.7 Anchor chain trench D/3 wide. V_{lug}-H_{lug} envelopes and governing ALS loads (Alderlieste et al., 2016).

TOTAL projects, each with an FPSO and off-loading terminal (OLT) buoy taut line mooring system. Instead of vertical sided trenches modelled previously, due to erosion and slope instability, the trenches are wider at the top than the base, with slope angles increasing to near-vertical for ≈ 3 m above the base. Figure 12.9a shows a cross-section 20 m away from an anchor pile ($D \approx 4.5$ m, $L \approx 17$ m).

Sassi et al. (2018) performed preliminary small scale (75 g–80 g) centrifuge tests on $D = 4.5$ m, $L = 15$ m, $z_{lug} = 9.5$ m suction piles jacked into either blocks extracted from undisturbed core samples ($s_u = 1 + 1.4$ z), or reconstituted clay (slightly higher s_u profile, gradients between 1.4 kPa/m and 2.0 kPa/m. A spatula, base width $= D = 4.5$ m (Figure 12.9b) excavated large trenches reaching the suction anchor and lug level depth. From the first series of tests (pure vertical pull-out; to model no reverse end-bearing (REB) the anchor tip was in contact with a sand layer), the presence of a trench had only a minor reduction (as expected). The "with REB" Q_{pl} ($= F_o + Q_{base}$) and "without REB" Q_{un} ($= F_o + F_i$) resistances were ≈ 3.0 MN and ≈ 1.6 MN respectively. Back-figured average α and $N_{c,circle}$ values were low (0.3, 8).

(a)

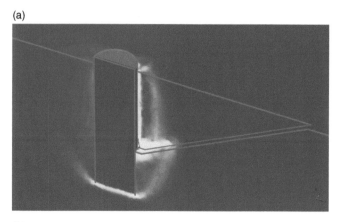

(b)

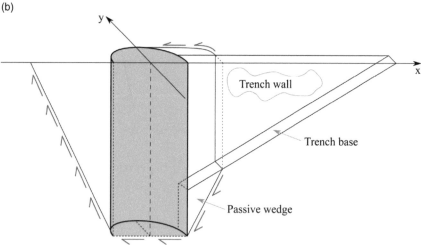

Figure 12.8 Anchor chain trench – "Fixed Head" incremental plastic strains and failure mechanism (Alderlieste et al., 2016).

For the remaining two test series, soil pore pressure data were obtained at ≈ 0.2 D, 1.2 D and 1.4 D below the tip, and these data confirmed that passive suction (i.e. REB) was developed even when trenches were present. Anchor holding capacity reductions of 11% were obtained for θ_{lug} between 30° and 40°. Surprisingly, this value is less than FEA by Hernandez-Martinez et al. (2015) and Arslan et al. (2015) (20% and 45%) who had 1D wide trenches and *included* full REB, but similar to that of Alderlieste et al. (2016) (≈ 7%) who had a smaller trench width (D/3) and *excluded* REB (i.e. they had a "plugged" soil response). 3D FEA predictions type C1 (Lambe, 1973) were also made.

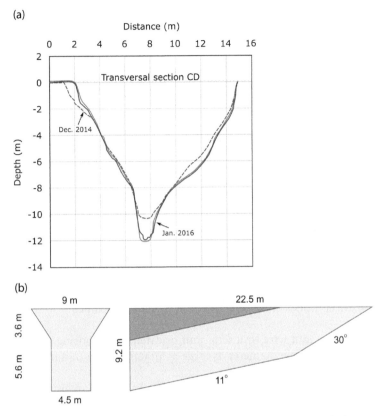

Figure 12.9 Observed anchor chain trench geometry in Gulf of Guinea clay (Colliat et al., 2018; Sassi et al., 2018).

Feng et al. (2019) presented results of a parametric 3D finite-element analyses investigation examining the geotechnical capacity of suction caissons in a trenching seabed with soil properties and caisson geometry reflecting the Serpentina field case. The results show that the reduction in the geotechnical capacity becomes more significant with increasing trench width due to the loss of soil support and a change in failure mechanism as the caisson rotates into the trench. For a given trench width, the reduction in capacity becomes more significant as the load inclination angle to the horizontal decreases. They show that a strategy to design for inevitable trenching by moving the lug shallower to reduce the depth of trench formation is not straightforward. The gain from a shallower trench may often be outweighed by the reduction in capacity from rotation of the caisson at failure for loading angles typical of taut moorings.

12.12.1 Commentary

Reverse end-bearing capacity accounts for $\approx$ 50% of anchor axial pull-out resistance. Excluding reverse end-bearing capacity adds something of the order of 2 m length for a 5 m diameter anchor pile (see the Fixed Head Pile Example in Section 10.7).

To date, most of the aforementioned work has been confined to the Gulf of Guinea (GoG). It is hoped that this will eventually extend to other areas of the world where anchor chain trenching has been observed.

One possible explanation for vertical trenches is that the Gulf of Guinea clays may be similar to those at Bothkennar UK. Both have strong in-situ structure, higher than usual sensitivity, and are lightly overconsolidated. Bothkennar soils have been subject to bioturbation by worms and exhibit rapid strain-softening due to destructuring (Atkinson et al., 1992; ICE, 1992). It is possible that vane shear test, T-bar and CPT testing all underestimate Gulf of Guinea clay peak undrained shear strength, $s_{u,peak}$. This may explain why vertical trench heights have been observed exceeding those given by classical wedge theory. (Mini) pressuremeter testing, plus a strain-softening material model (Kay and Van Woensel, 2002) may be useful.

Mitigation measures for anchor chain trenching include:

- replacing taut wire by a semi-taut or catenary anchoring system
- making the pile diameter as large as practicable to minimise capacity decrease
- putting the lug further below optimum lug level (say 1 m instead of 0.5 m) to decrease the risk of forward rotation
- designing to exclude reverse end-bearing in order to account for the possible risk of water ingress (i.e. adopt the Figure 10.1 "leaking" failure mode).

12.13 CLOSURE

This ultimate chapter has covered miscellaneous geotechnical design considerations. Purely structural considerations have been omitted, for example pile tip integrity and design of offshore wind turbine monopile transition piece connections.

The breadth of the topics, which is greater than for either shallow or deep foundations, emphasizes the point made in Section 8.1.1 that "intermediate foundation design usually requires close cooperation between geotechnical and other offshore disciplines, particularly structural engineers".

Acronyms

AG	Fugro Advanced Geomechanics P/L, Australia
aka	Also Known As
ALS	Accidental Limit State
API	American Petroleum Institute, USA
ASTM	American Society for Testing and Materials, USA
BE	Best Estimate
bsf	Below Sea Floor
CD	Consolidated Drained
CEL	Coupled Eulerian Lagrangian
CEN	Comité Européen de Normalisation
COFS	Centre for Offshore Foundation Systems, University of Western Australia (UWA)
COV	Coefficient of Variation
CPT	Cone Penetration Test
CPTu	Cone Penetration Test with pore pressure measurement
CU	Consolidated Undrained
DNV-GL	Det Norske Veritas – Germanischer Lloyd A/S, Norway
DOF	Degree of Freedom
DSS	Direct Simple Shear
EDF-EN	Electricité de France – Energies Nouvelles, France
FEA	Finite Element Analysis
FEBV	Fugro Engineers BV, the Netherlands
FLS	Fatigue Limit State
FOS	Factor of Safety
FPSO	Floating Production Storage and Off-loading
GoG	Gulf of Guinea
GoM	Gulf of Mexico
HE	High Estimate
IEC	International Electrical Commission
ISFOG	International Symposium on Frontiers in Offshore Geotechnics
ISO	International Organization for Standardization
ISOPE	International Society of Offshore and Polar Engineers
JIP	Joint Industry Project

LAT	Lowest Astronomical Tide
LC	Load Case
LDFE	Large Deformation Finite Element
LE	Low Estimate
LOC	Lightly Over Consolidated
LRFD	Load and Resistance Factor Design
MODU	Mobile Offshore Drilling Unit
NC	Normally Consolidated
NGI	Norwegian Geotechnical Institute, Norway
OC	Over Consolidated
OLL	Optimum Lug Level
OLT	Offloading Terminal (i.e. offshore buoy)
OMAE	Offshore Mechanics and Arctic Engineering
OTC	Offshore Technology Conference, USA
OTRC	Offshore Technology Research Center, USA
OWT	Offshore Wind Turbine
PFD	Partial Factor Design
PLEM	Pipe Line End Manifold
PLET	Pipe Line End Termination
PMT	Pressuremeter Test
PSD	Particle Size Distribution
REB	Reverse End-Bearing
RP	Reference Point (loads and displacements)
RQD	Rock Quality Designation
RTA	Riser Tower Anchor
SDOF	Single Degree of Freedom
SLS	Serviceability Limit State
SPT	SPT Offshore BV, The Netherlands
SRD	Soil Resistance to Driving
SUT	Society for Underwater Technology, UK
SWP	Self-Weight Penetration
TPD	Target Penetration Depth
UCS	Unconfined Compressive Strength
ULS	Ultimate Limit State
UU	Unconsolidated Undrained
WD	Water Depth
WHP	Well Head Platform
WT	Wall Thickness

Notation

The terms caisson and foundation, lug and padeye are synonymous.

$\alpha_{D,su}$ = s_u averaging parameter ($\alpha_{D,su} \approx 0.25$)
= clay undrained strength averaged from L to L + D $\alpha_{D,su}$

α_i = clay-steel inner adhesion factor (unit friction $f_i = \alpha_i\ s_u$)

α_o = clay-steel outer adhesion factor (unit friction $f_o = \alpha_o s_u$)

α_u = excess pore pressure ratio at tip level = $\Delta u(tip)/\Delta u$

β_i = sand-steel inner adhesion factor (unit friction $f_i = \beta_i\ \sigma'_{vo}$)

β_o = sand-steel outer adhesion factor (unit friction $f_o = \beta_o\ \sigma'_{vo}$)

γ_{sub} = soil submerged unit weight

γ_{water} = (sea) water unit weight

δ_x = caisson head lateral displacement

δ_z = caisson head vertical displacement

ρ_{soil} = soil density

ρ_{steel} = steel density

ρ_{water} = (sea) water density

ΔH = vent (nozzle) differential head loss

Δu = under/overpressure

Δu_{all} = allowable under/overpressure

θ_{xz} = caisson rotation

θ_{lug} = anchor chain inclination from horizontal at lug level depth z_{lug}

θ_{z0} = anchor chain inclination from horizontal at seafloor

λ_L = partial load factor (API PFD/LRFD)

$\lambda_{L,Bw}$ = partial load factor on $B_{w,steel}$ and $B_{w,plug}$

φ_L = partial load factor (ISO 19902)

φ_m = partial resistance factor (API PFD/LRFD)

γ_m = partial resistance factor (ISO 19901)

σ_c = rock unconfined compressive strength

σ'_v = soil effective vertical stress

σ'_{vo} = soil effective in-situ vertical stress

Φ_{MH} = ellipse rotation angle
= $atan(e_{z,su}/L) - \Delta\Phi_{MH}$

$\Delta\Phi_{MH}$ = ellipse rotation angle correction

ν	= soil drained Poisson's ratio
ν_u	= soil undrained Poisson's ratio = 0.5
A_{base}	= caisson base area
	= $\pi D^2/4$
A_i	= caisson internal (suction) area
	= $\pi D_i^2/4$
A_p, A_{wall}	= caisson wall tip area
	= $\pi (D^2 - D_i^2)/4$
A_{si}	= caisson inner perimeter
	= πD_i
A_{so}	= caisson outer perimeter
	= πD
A_n	= vent (nozzle) area
a_{MH}	= ellipse major semi-axis length
b_{MH}	= ellipse minor semi-axis length
Δb_{MH}	= ellipse minor semi-axis length correction
a_{VH}	= ellipsoid parameter
b_{VH}	= ellipsoid parameter
C_d	= vent (nozzle) discharge coefficient
D	= caisson outer diameter
D_i	= caisson inner diameter
	= $D - 2\,WT$
D_r	= (cohesionless) soil relative density [–]
E	= soil drained Young's Modulus
E_u	= soil undrained Young's Modulus
e_z	= H load vertical eccentricity with respect to caisson head (seafloor) = M/H
$e_{z,su}$	= analytical e_z based on $s_{u,z} = \int_0^L s_{u,z}\, z\, dz \,/ \int_0^L s_{u,z}\, dz$
$e_{z,Hmax}$	= e_z based on M/H at H_{max}
F_i	= caisson inner skin friction resistance
F_o	= caisson outer skin friction resistance
g	= acceleration due to gravity
FOS	= overall factor of safety
H	= horizontal load at caisson head (seafloor)
H^*	= non-dimensional H value = $H/(D\,L\,s_{u,av,L})$
H^{**}	= non-dimensional H value = $H/(D^2\,s_{u,av,L})$
H_{max}	= caisson maximum "fixed head" horizontal resistance (V = 0)
	= $N_{p,fixed}\,L\,D\,s_{u,av,L}$
$H_{max,V}$	= as H_{max}, but with $V \neq 0$
H_o	= "free head" horizontal resistance (V and M = 0) = $N_{p,free}\,L\,D\,s_{u,av,L}$
i_{avg}	= soil plug average upwards seepage gradient (= head loss/L)
i_{crit}	= soil plug critical upwards seepage gradient (= γ_{sub}/γ_w)
k_f	= empirical coefficient relating q_c (or q_t) to unit skin friction resistance

k_p = empirical coefficient relating q_c (or q_t) to unit end bearing resistance

k_r = sand plug : soil mass permeability ratio

L = caisson embedded length

L_{plug} = soil plug length (usually > L)

$(L/D)_{crit}$ = caisson critical embedded length : diameter ratio

M = moment load at caisson head (seafloor)

M^* = non-dimensional M value = $M/(D L^2 s_{u,av,L})$

M^{**} = non-dimensional M value = $M/(D^2 L s_{u,av,L})$

M_{max} = caisson maximum moment resistance (V = 0)

M_o = caisson moment resistance (V and H = 0)

$N_{c,strip}$ = strip foundation bearing capacity factor (clay) $\approx$ min[5 (1 + 0.2 L/D), 7.5]

$N_{c,circle}$ = circular foundation bearing capacity factor (clay) $\approx$ min[6 (1 + 0.2 L/D), 9]

$N_{p,fixed}$ = "fixed head" lateral bearing capacity factor (clay) = $H_{max}/(D L s_{u,av,L})$

$N_{p,free}$ = "free head" lateral bearing capacity factor (clay) = $H_o/(D L s_{u,av,L})$

N_q = dimensionless bearing capacity factor (sand)

$q_c (q_T)$ = CPT cone (total) tip resistance

$q_{c,z}$ = CPT cone tip resistance at depth z below seafloor

$q_{c,L}$ = CPT cone tip resistance at caisson embedded/tip depth L

R_n = vent (nozzle) resistance

r_{lug} = lug radial offset from centreline

S_t = soil sensitivity = s_u/s_{ur}

$s_{u,z}$ = soil undrained shear strength at depth z

s_{ur} = soil remoulded (residual) undrained shear strength

$s_{u,0}$ = soil undrained shear strength at seafloor (z = 0)

$s_{u,av,L}$ = average $s_{u,z}$ between caisson head (seafloor) and caisson tip (L)

$s_{u,tip}$ = soil undrained shear strength at foundation tip depth (L)

$s_{u,av,tip}$ = average $s_{u,z}$ between caisson tip (L) and depth (L + D $\alpha_{D,su}$)

$s_{u,C}$ = soil undrained shear strength in triaxial compression

$s_{u,DSS}$ = soil undrained shear strength in direct simple shear

$s_{u,E}$ = soil undrained shear strength in triaxial extension

t = MH resistance ellipse parameter (0 < t < 2π)

T = torsion load at caisson head (seafloor)

T_{max} = caisson maximum torsional resistance (V, H and M = 0)

T_{lug} = anchor chain load at caisson lug level (z_{lug})

T_{z0} = anchor chain load at seafloor

v_0 = foundation (free-fall) velocity at seafloor

v_{pile} = foundation (free-fall) velocity

v_n = vent (nozzle) velocity

V = vertical load at caisson head (seafloor)

V_{max} = caisson axial resistance (T, H and M = 0)

W_{plug} = soil plug submerged weight

W_{steel} = caisson submerged weight
WT = caisson wall thickness
WT_{top} = caisson top plate thickness
z = depth below seafloor
z_{lug} = lug level depth below seafloor
$z_{lug,cL}$ = lug level depth below seafloor on caisson centreline

References

Abadie, C.N., Byrne, B.W. and Houlsby, G.T. (2017) *Modelling of monopile response to cyclic lateral loading in sand. Proc. of the 8th Int. Conf. Offshore Site Investigation and Geotechnics.* OSIG 2017, London, Society of Underwater Technology, pp. 1046–1053.

Abbs, A.F. (1983) *Lateral pile analysis in weak carbonate rocks. Proc. Conference on Geotechnical Practice in Offshore Engineering: University of Texas at Austin,* Austin, Texas, USA, 27–29 April, pp. 546–556.

Abbs, A.F. and Needham, A.D. (1985) *Grouted piles in weak carbonate rocks. Proc. Offshore Technology Conference,* Houston, Texas, USA, Paper OTC-4852-MS.

Achmus, M., Kuo, Y-S. and Abdel-Rahman, K. (2009) Behavior of monopile foundations under cyclic lateral load. *Computers and Geotechnics,* Vol. 36, pp. 725–735.

Advanced Geomechanics (2002) *Suction Pile Analysis Code, AGSpanc Version 4.2, User's Manual.* Doc. No. AGR-1176, Rev 5.

AG (2003) *AGSPANC: Suction Pile Analysis Code,* User Manual Version 4.2 Advanced Geomechanics, Australia.

AGS (2000) *Guidelines for Combined Geoenvironmental and Geotechnical Investigations,* Association of Geotechnical and Geoenvironmental Specialists, Beckenham, UK.

Ahn, J., Lee, H. and Kim, Y-T. (2014) Holding capacity of suction caisson anchors embedded in cohesive soils based on finite element analysis. *International Journal for Numerical and Analytical Methods in Geomechanics,* Vol. 38, no. 15, pp. 1541–1555.

Alderlieste, E.A. and Dekker, M.J. (2018) *Suction pile design and installation challenges for the Ophir WHP. Proc. 1st Vietnam Symposium on Advances in Offshore Engineering (VSOE 2018),* Hanoi, Vietnam, Springer, pp. 329–335.

Alderlieste, E.A., Romp, R.H., Kay, S. and Lofterød, A. (2016) *Assessment of seafloor trench for suction pile moorings: a field case. Proc. Offshore Technology Conference,* Houston, Texas, USA, Paper OTC-27035-MS.

Alderlieste, E.A. and van Blaaderen, E.A. (2015) *Installation of suction caissons for an asymmetrical support structure in sandy soil. Proc. 3rd International Symposium on Frontiers in Offshore Geotechnics,* Taylor & Francis, Oslo, Norway, pp. 215–220.

Aldridge, T.R., Carrington, T.M. and Kee, N.R. (2005) *Propagation of pile tip damage during installation*. Proc. 1st International Symposium on Frontiers in Offshore Geotechnics, Taylor & Francis, Perth, Australia, pp. 823–827.

Andersen, K.H. (2015) *Cyclic soil parameters for offshore foundation design*. Proc. 3rd International Symposium Frontiers in Offshore Geotechnics, June, Oslo, Norway, pp. 5–82.

Andersen, K.H. and Jostad, H.P. (1999) *Foundation design of skirted foundations and anchors in clay*. Proc. Offshore Technology Conference, Houston, Texas, USA, Paper OTC-10824-MS.

Andersen, K.H. and Jostad, H.P. (2004) *Shear strength along inside of suction anchor skirt wall in clay*. Proc. Offshore Technology Conference, Houston, Texas, USA, Paper OTC-16844-MS.

Andersen, K.H., Jostad, H.P. and Dyvik, R. (2008) Penetration resistance of offshore skirted foundations and anchors in dense sand. *Journal of Geotechical and Geoenvironmental Engineering*, Vol. 134, no. 1, pp. 106–116.

Andersen, K.H., Murff, J.D., Randolph, M.F., Clukey, E.C., Erbrich, C.T., Jostad, H.P., Jansen, B., Aubeny, C., Sharma, P. and Supachawarote, C. (2005) *Suction anchors for deepwater applications*. Proc. 1st International Symposium on Frontiers in Offshore Geotechnics, Taylor & Francis, Perth, Australia, pp. 3–30.

API (1997) Recommended practice for planning, designing and constructing Tension Leg Platforms RP 2T, Ed 2 American Petroleum Institute, Washington, DC. (Ed 3 2015).

API (2000) Recommended practice for planning, designing and constructing fixed offshore platforms - working stress design RP-2A, American Petroleum Institute, Washington, USA.

API (2005) Recommended practice for design and analysis of station keeping systems for floating structures, API RP-2SK, American Petroleum Institute, Washington, USA.

API (2008) Recommended practice for design and analysis of station keeping systems for floating structures, API RP-2SK, American Petroleum Institute, Washington, USA.

API (2011) Recommended practice for geotechnical and foundation design considerations RP 2GEO, Ed 1 American Petroleum Institute, Washington, DC. Errata (2014).

Arany, L., Bhattacharya, S., Macdonald, L. and Hogan, S.J. (2017) Design of monopiles for offshore wind turbines in 10 steps. *Soil Dynamics and Earthquake Engineering*, Vol. 92, pp. 126–152.

ARGEMA (1994) Foundations in Carbonate Soils, English edn, Technip, Paris.

Arslan, H., Peterman, B.R., Wong, P.C. and Bhattacharjee, S. (2015) *Remaining capacity of the suction pile due to seabed trenching*. Proc. 25th International Symposium on Ocean and Polar Engineering Conference, Hawaii, pp. 924–931.

ASTM (2009) Standard practice for description and identification of soils (Visual-Manual Procedure), ASTM D2488-09a. USA.

ASTM (2011) Standard practice for classification of soils for engineering purposes (Unified Soil Classification System), ASTM D2487-11. USA.

Atkinson, J.H., Allman, M.A. and Böese, R.J. (1992) Influence of laboratory sample preparation procedures on the strength and stiffness of intact Bothkennar soil recovered using the Laval sampler. *Géotechnique*, Vol. 42, no. 2, pp. 349–354.

Attewell, P.B. and Farmer, I.W. (1976) *Principles of engineering geology*, Chapman and Hall, London.

Baguelin, F., Frank, R. and Saïd, Y.H. (1977) Theoretical study of lateral reaction mechanism of piles. *Géotechnique*, Vol. 27, no. 3, pp. 405–434.

Balder, T., de Lange, D.A., Elkadi, A.S.K., Egberts, P.J.P., Beuckelaers, W.J.A.P., Coronel, M., van Dijk, J., Atkinson, R. and Luger, H.J. (2020) *Hydraulic pile extraction scale tests (HyPE-ST): experimental design and preliminary results.* Proc. 4th International Symposium on Frontiers in Offshore Geotechnics, article #3484, publication #1069 (IC-ISFOG21), 8–11 November 2021 (original dates 16–19 August 2020), Austin, Texas, USA, ASCE Geo-Institute and DFI (Proc published in August 2020).

Ballard, J-C., Delvosal, O. and Yonatan, P.H. (2010) *Simplified VH equations for foundation punch-through sand into clay. Proc. 2nd International Symposium on Frontiers in Offshore Geotechnics*, Perth, Australia, pp. 650–655.

Barbosa, P. Geduhn, M. Jardine, R.J. Schroeder, F.C. and Horn, M. (2015) *Offshore pile load tests in chalk. Proc. 16th Eur. Conf. Soil Mech. Geotech. Eng*, Edinburgh, Scotland, pp. 2885–2890.

Bhattacharjee, S., Majhi, S.M., Smith, D. and Garrity, R. (2014) *Serpentina FPSO mooring integrity issues and system replacement: unique fast track approach. Proc. Offshore Technology Conference*, Houston, Texas, USA, Paper OTC-25449-MS.

Bransby, M.F. and Yun, G-J. (2009) Undrained capacity of skirted strip foundations under combined loading. *Géotechnique*, Vol. 59, no. 2, pp. 115–125.

Broms, B.B. (1964a) Lateral resistance of piles in cohesive soils. *Journal of the Soil Mechanics and Foundations Division, Proc. American Society of Civil Engineers*, Vol. 90, no. SM2, pp. 27–63.

Broms, B.B. (1964b) Lateral resistance of piles in cohesionless soils. *Journal of the Soil Mechanics and Foundations Division, Proc. American Society of Civil Engineers*, Vol. 90, no. SM3, pp. 123–156.

Broughton, P., Davies, R.L., Aldridge, T. and Carrington, T. (2002) Foundation design for the refloat of the Maureen steel gravity platform. *Proc. ICE, Geotechnical Engineering Journal*, Vol. 155, no. 2, pp. 111–118.

Brown, G.A. and Nacci, V.A. (1971) *Performance of hydrostatic anchors in granular soils. Proc. Offshore Technology Conference*, Houston, Texas, USA, Paper OTC-1472-MS.

BSI (1999) *Code of practice for site investigations*, British Standard BS 5930:1999. British Standards Institution, UK.

Burd, H., Byrne, B., McAdam, R., Housby, G., Martin, C., Beuckelaers, W., Zdravkovic, L., Taborda, D., Potts, D., Jardine, J., Gavin, K., Doherty, P., Igoe, D., Skov Gretlund, J., Pacheco Andrade, M. and Muir Wood, A. (2017) *Design aspects for monopile foundations. Proc. Workshop TC209 19th ICSMGE*, Seoul, 2017, pp. 35–44.

Burd, H., David, M.G., Taborda, D., Zdravkovic, L., Abadie, C., Byrne, B., Housby, G., Gavins, K.G., Igoe, D.J.P., Jardine, R.J., Martin, C., McAdam, R., Pedro, A.M.G. and Potts, D.M. (2020) PISA design model for monopiles for offshore wind turbines: application to a marine sand. *Géotechnique*. doi:10.1680/jgeot.18.P.277.

Burt, N.J. and Harris, R.P. (1980) *Design, installation and testing of belled pile foundations. Proc. Offshore Technology Conference*, Houston, Texas, USA, Paper OTC-3872-MS.

Butterfield, R. and Gottardi, G. (1994) A complete three-dimensional failure envelope for shallow footings on sand. *Géotechnique*, Vol. 44, no. 1, pp. 181–184.

Bye, A., Erbrich, C., Rognlien, B. and Tjelta, T.I. (1995) *Geotechnical design of bucket foundations. Proc. Offshore Technology Conference*, Houston, Texas, USA, Paper OTC-7793-MS.

Byrne, B., Housby, G., Burd, H., Gavin, K.G., Igoe, D.J.P., Jardine, R.J., Martin, C., McAdam, R., Potts, D.M., David, M.G., Taborda, D., Zdravkovic, L. (2020) PISA design model for monopiles for offshore wind turbines: application to a stiff glacial clay till. *Géotechnique*. Published Online doi:10.1680/jgeot.18.P.255.

Byrne, B.W. (2017) *New design methods for offshore wind turbine monopiles. Proc 8th International Conference on Offshore Site Investigation and Geotechnics*, held 12–14 September 2017, Royal Geographical Society, London.

Byrne, B.W., McAdam, R., Burd, H.J., Houlsby, G.T., Martin, C.M., Zdravkovic, L., Taborda, D.M.G., Potts, D.M., Jardine, R.J., Sideri, M., Schroeder, F.C., Gavin, K., Doherty, P., Igoe, D., Muir Wood, A., Kallehave, D. and Skov Gretlund, J. (2015a) *New design methods for large diameter piles under lateral loading for offshore wind applications. Proc. 3rd International Symposium Frontiers in Offshore Geotechnics*, June, Oslo, Norway, pp. 705–710.

Byrne, B.W., McAdam, R., Burd, H.J., Houlsby, G.T., Martin, C.M., Gavin, K., Doherty, P., Igoe, D., Zdravkovic, L., Taborda, D.M.G., Potts, D.M., Jardine, R.J., Sideri, M., Schroeder, F.C., Muir Wood, A., Kallehave, D. and Skov Gretlund, J. (2015b) *Field testing of large diameter piles under lateral loading for offshore wind applications. Proc. XVI ECSMGE Geotechnical Engineering for Infrastructure and Development*, Edinburgh, Scotland, pp. 1255–1260.

Carbon Trust (2019) *Suction installed caisson foundations for offshore wind: design guidelines*. Carbon Trust. https://prod-drupal-files.storage.googleapis.com/documents/resource/public/owa-suction-caisson-design-guidelines-report.pdf.

Carter, J.P. and Kulhawy, F.H. (1992) Analysis of laterally loaded shafts in rock. *ASCE Journal of Geotechnical Engineering*, Vol. 118, no. 6, pp. 839–855.

Cathie Associates (2014) *OPILE instruction manual*, Cathie Associates SA/NV, Belgium.

CEN (2004) *Eurocode 7: geotechnical design - part 1: general rules*, European Standard EN 1997-1:2004. (With Corrigendum EN 1997-1:2004/AC, February 2009, European Committee for Standardization.

Cerfontaine, B., Collin, F. and Charlier, R. (2016) Numerical modelling of transient cyclic vertical loading of suction caissons in sand, *Géotechnique*, Vol. 66, no. 2, pp. 121–136.

CFMS (2019) *Recommendations for planning and designing foundations of offshore wind turbines*, French workgroup 'Foundations of Offshore Wind Turbines, Version 2.1. http://www.cfms-sols.org/sites/default/files/Rapport-cfms-eoliennes-offshore-EN.pdf.

Chan, S.F. and Hannah, T.H. (1980) Repeated loading on single piles in sand. *ASCE Journal of Geotechnical and Geoenvironmental Engineering*, Vol. 106, no. GT2, pp. 171–188.

Chatzivasileiou, I. (2014) *Installation of suction caissons in layered sand*, MSc. Thesis, Delft University of Technology, The Netherlands.

Chen, W. and Randolph, M.F. (2007a) External radial stress changes and axial capacity for suction caisson in soft clay. *Géotechnique*, Vol. 57, no. 6, pp. 499–511.

Chen, W. and Randolph, M.F. (2007b) Uplift capacity of suction caissons under sustained and cyclic loading in soft clay. *ASCE Journal of Geotechnical and Geoenvironmental Engineering*, Vol. 133, no. 11, pp. 1352–1363.

Chow, Y.K. (1981) *Dynamic behaviour of piles*. Ph.D. Thesis, University of Manchester, UK.

CIRIA (1984) Report 103 *Design of laterally-loaded piles*, CIRIA, London, UK, CIRIA Report 103. Elson, W.K., Construction Industry Research and Information Association.

CIRIA (2002) *C574 Engineering in chalk*, Lord, J. A. Clayton, C. R. I. and Mortimore, R. N., Construction Industry Research and Information Association, UK.

Clausen, C.J.F. and Tjelta, T.I. (1996) *Offshore platforms supported by bucket foundations. 15th Congress, Copenhagen, June 16–20: Congress report, international association for bridge and structural engineering IABSE*, Zürich, pp. 819–829.

Clukey, E.C. and Morrison, M.J. (1993) A centrifuge and analytical study to evaluate suction caissons for TLP applications in the Gulf of Mexico. *Design and Performance of Deep Foundations, Piles and Piers in Soil and Soft Rock*, ASCE Geotechnical Special Publication GSP No. 38, pp. 141–156.

Clukey, E.C., Morrison, M.J., Garnier, J. and Corté, J.F. (1995) *The response of suction caissons in normally consolidated clays to cyclic TLP loading conditions. Proc. Offshore Technology Conference*, Houston, Texas, USA, Paper OTC-7796-MS.

COFS (2003) *Deepwater anchor design practice. Phase II Report to API/Deepstar*. Vol. II-2 Appendix B. COFS 3D FE Capacity Analyses of Suction Caisson Anchors. Australia.

Colliard, D. and Wallerand, R. (2008) *Design and installation of suction piles in West Africa deepwaters. Proc. 2nd British Geotechnical Association International Conference on Foundations*, 24–27 June, University of Dundee, Scotland, UK, IHS BRE Press, pp. 825–836.

Colliat, J.-L. and Dendani, H. (2002) *Girassol: geotechnical design analyses and installation of the suction anchors. Proc. Society for Underwater Technology Conf. Offshore Site Investigation and Geotechnics*, London, pp. 107–122.

Colliat, J-L., Safinus, S., Boylan, N. and Schroeder, K. (2018) *Formation and development of seabed Trenching from subsea inspection data of deepwater Gulf of Guinea moorings. Proc. Offshore Technology Conference*, Houston, Texas, USA, Paper OTC-29034-MS.

Cotter, O. (2009) *The installation of suction caisson foundations for offshore renewable energy structures*, PhD. Thesis, Oxford, UK.

Cox, W.R., Dixon, D.A. and Murphy, B.S. (1984) *Lateral load tests on 25.4-mm (1-in.) diameter piles in very soft clay in side-by-side and in-line groups. Laterally loaded deep foundations: analysis and performance*, ASTM STP 835, American Society for Testing and Materials, Kansas City, Missouri, USA, pp. 122–139.

De Araujo, J.B., Machado, R.D. and Medeiros, C.J. (2004) *High holding power torpedo pile—results for the first long term application. Proc., 23rd Int. Conf. on Offshore Mechanics and Arctic Engineering.* ASME, Paper No. OMAE2004-51201.

Degenkamp, G. and Dutta, A. (1989) Soil resistances to embedded anchor chain in soft clay. *Journal of Geotechnical Engineering*, Vol. 115. doi:10.1061/(ASCE)0733-9410(1989)115:10(1420).

Dekker, M.J. (2014) *The modelling of suction caisson foundations for multi-footed structures*, MSc. Thesis, Delft University of Technology, The Netherlands & Norwegian University of Science and Technology (NTNU), Trondheim, Norway.

Dendani, H. (2003) *Suction anchors: some critical aspects for their design and installation in clayey soils. Proc. Offshore Technology Conference*, Houston, Texas, USA, Paper OTC-15376-MS.

Devriendt, C., Van Ingelgem, Y., De Sitter, G., Jordaens, P.J. and Guillaume, P. (2013) *Monitoring of resonant frequencies and damping values of an offshore wind turbine on a monopile foundation*, Offshore Wind Infrastructure Application Lab, Belgium, presentation to EWEA 2013.

Dijkhuizen, C., Coppens, T, and van der Graaf, P. (2003) *Installation of the Horn Mountain Spar using the enhanced DCV Balder. Proc. Offshore Technology Conference*, Houston, Texas, USA, Paper OTC-16367-MS.

DNV (1992) *Foundations*, Classification Notes No. 30.4. Det Norske Veritas.

DNV (2008) *Design of Offshore Steel Structures*, General (LRFD Method), Offshore Standard DNV-OS-C101. Det Norske Veritas.

DNV (2014a) *Design of Offshore Wind Turbine Structures*, Offshore Standard DNV-OS-J101. Det Norske Veritas.

DNV (2014b) *Modelling and Analysis of Marine Operations*, Recommended Practice DNV-RP-H103. Det Norske Veritas.

DNV (2018) *Geotechnical Design and Installation of Suction Anchors in Clay*, Recommended Practice DNV-RP-E303. Det Norske Veritas.

DNV-GL (2016) *Support Structures for Wind Turbines*, Standard DNVGL-ST-0126.

DNV-GL (2017, 2019) *Offshore Soil Mechanics and Geotechnical Engineering*, Recommended Practice DNVGL-RP-C212.

Doherty, J.P. and Deeks, A.J. (2003) Elastic response of circular footings embedded in a nonhomogeneous half-space. *Géotechnique*, Vol. 53, no. 8, pp. 703–714.

Doherty, J.P., Houlsby, G.T. and Deeks, A.J. (2005) Stiffness of flexible caisson foundations embedded in nonhomogeneous elastic soil. *ASCE Journal of Geotechnical and Geoenvironmental Engineering*, Vol. 131, no. 12, pp. 1498–1508.

Dührkop, J. and Grabe, J. (2009) *Design of laterally loaded piles with bulge. Proc. 28th International Conference on Ocean, Offshore and Arctic Engineering*, ASME, Honolulu, Hawaii, USA. OMAE2009-79087.

Dührkop, J., Maretzki, S. and Rieser, J. (2017) *Re-evaluation of pile driveability in chalk. Proc. of the 8th Int. Conf. Offshore Site Investigation and Geotechnics.* OSIG 2017, Society of Underwater Technology, London, Vol. 2, pp. 666–673.

Dyson, G.J. and Randolph, M. (2001) Monotonic lateral loading of piles in calcareous sediments. *ASCE Journal of Geotechnical and Geoenvironmental Engineering*, Vol. 127, no. 24, pp. 346–352.

Edwards, D.H., Zdravkovic, L. and Potts, D.M. (2005) Depth factors for undrained bearing capacity. *Géotechnique*, Vol. 55, no. 10, pp. 755–758.

Engin, H.K., Khoa, H.D.V., Jostad, H.P. and Alm, T. (2015) *Finite element analyses of spudcan – subsea template interaction during jack-up rig installation. Proc. 3rd International Symposium Frontiers in Offshore Geotechnics*, June, Oslo, Norway, 705, pp. 1275–1280.

Erbrich, C.T. (2004) *A new method for the design of laterally loaded anchor piles in soft rock. Proc. Offshore Technology Conference*, Houston, Texas, USA, Paper OTC-16441-MS.

Erbrich, C.T., Barbosa-Cruz, E. and Barbour, R. (2010) *Soil-pile interaction during extrusion of an initially Deformed Pile. Proc. 2nd International Symposium on Frontiers in Offshore Geotechnics*, Perth, Australia, pp. 489–494.

Erbrich, C.T. and Hefer, P. (2002) *Installation of the Laminaria suction piles – A case history. Proc. Offshore Technology Conference*, Houston, Texas, USA, Paper OTC-14220-MS.

Erbrich, C.T., Lam, S.Y., Zhu, H., Derache, A., Sato, A. and Al-Showaiter, A. (2017) *Geotechnical design of anchor piles for Ichthys CPF and FPSO. Proc. of the 8th Int. Conf. Offshore Site Investigation and Geotechnics.* OSIG 2017, Society of Underwater Technology, London, pp. 1186–1197.

Erbrich, C.T. and Tjelta, T.I. (1999) *Installation of bucket foundations and suction caissons in sand – geotechnical performance. Proc. Offshore Technology Conference,* Houston, Texas, USA, Paper OTC-10990-MS.

Feld, T. (2001) *Suction buckets: a new innovation foundation concept, applied to offshore wind turbines,* PhD. Thesis, Aalborg University, Denmark.

Feng, X., Gourvenec, S. and White, D.J. (2019) Load capacity of caisson anchors exposed to seabed trenching. *Ocean Engineering,* Vol. 171, no. 1, pp. 181–192.

Feng, X., Randolph, M.F., Gourvenec, S. and Wallerand, R. (2014) Design approach for rectangular mudmats under fully three dimensional loading. *Géotechnique,* Vol. 64, no. 1, pp. 51–63.

Finnie, I.M.S. and Randolph, M.F. (1994) *Punch-through and liquefaction induced failure of shallow foundations on calcareous sediments. Proc. 7th International Conference on the Behaviour of Offshore Structures,* BOSS '94, Vol. 1, Pergamon, Oxford, pp. 217–230.

Fisher, R. and Cathie, D. (2003) *Optimisation of gravity based design for subsea applications. Proc. 2nd British Geotechnical Association International Conference on Foundations,* 2–5 September, University of Dundee, Scotland, UK, IHS BRE Press.

Fleshman, M.S. and Rice, J.D. (2014) Laboratory modeling of the mechanisms of piping erosion initiation. *ASCE Journal of Geotechnical and Geoenvironmental Engineering,* Vol. 140, no. 6. doi:10.1061/(ASCE)GT.1943-5606.0001106.

Fragio, A.G., Santiago, J.L. and Sutton, V.J.R. (1985) *Load tests on grouted piles in rock. Proc. Annual Offshore Technology Conference,* Houston, Texas, USAs, Paper OTC-4851-MS.

Frankenmolen, S.F., Erbrich, C.T. and Fearon, R.E. (2017) *Successful installation of large suction caissons and driven piles in carbonate soils. Proc. of the 8th Int. Conf. Offshore Site Investigation and Geotechnics.* OSIG 2017, London, Society of Underwater Technology, pp. 539–548.

Fuglsang, L.D. and Steensen-Bach, J.O. (1991) *Breakout resistance of suction piles in clay. Centrifuge 91: Proc. International conference centrifuge,* Boulder, Colorado, 13–14 June 1991, A.A. Balkema, Rotterdam, pp. 153–159.

Fugro (2009) *CANCAP2 User's Manual,* Fugro Engineers BV Internal Document No. FEBV/CDE/MAN/083.

Gazetas, G. (1991) Foundation vibrations. In: *Foundation engineering handbook,* Fang, H-Y. (Ed.), 2nd edn, Van Nostrand Reinhold Company, New York, pp. 553–593.

Germanischer Lloyd (2013) *Rules for the certification and construction, IV Industrial services,* Part 7 Offshore substations, Chapter 2 Structural Design, Germanischer Lloyd, Hamburg.

Germanischer Lloyd Wind Energie GmbH (2005) *Rules and guidelines: IV – Industrial services, Part 2 – Guideline for the certification of offshore wind turbines,* Germanischer Lloyd, Hamburg.

Gerolymos, N., Zafeirakos, A. and Karapiperis, K. (2015) Generalized failure envelope for caisson foundations in cohesive soil: static and dynamic loading. *Soil Dynamics and Earthquake Engineering,* Vol. 78, pp. 154–174.

Gilbert, R.B. and Murff, J.D. (2001) *Identifying uncertainties in the design of suction caisson foundations. Proc. International Conference on Geotechnical, Geological and Geophysical Properties of Deepwater Sediments honoring Wayne A. Dunlap*, OTRC, Houston, Texas, pp. 231–242.

Goodman, L.J., Lee, C.N. and Walker, F.J. (1961) The feasibility of vacuum anchorage in soil. *Géotechnique*, Vol. 1, no. 3, pp. 356–359.

Gottardi, G., Houlsby, G.T. and Butterfield, R. (1999) Plastic response of circular footings on sand under general planar loading. *Géotechnique*, Vol. 49, no. 4, pp. 453–469.

Gourvenec, S. and Jensen, K. (2009) Effect of embedment and spacing on co-joined skirted foundation systems on undrained limit states under general loading. *International Journal of Geomechanics*, Vol. 9, no. 6, pp. 267–279.

Gourvenec, S. and Steinepreis, M. (2007) Undrained limit states of shallow foundations acting in consort. *ASCE, International Journal of Geomechanics*, Vol. 7, no. 3, pp. 194–205.

Gourvenec, S.M., Feng, X., Randolph, M.F. and White, D.J. (2017) *A toolbox for optimizing geotechnical design of subsea foundations. Proc. Annual Offshore Technology Conference*, Houston, Texas, USA Paper OTC-27703-MS.

Govoni, L, Gottardi, G., Antoninia, A., Archettia, R. and Schweizer, J. (2016) Caisson foundations for competitive offshore wind farms in Italy. *Procedia Engineering*, Vol. 158, pp. 392–397.

Griffiths, D.V. (1982) Elasto-plastic analyses of deep foundations in cohesive soil. *International Journal for Numerical and Analytical Methods in Geomechanics*, Vol. 6, pp. 211–218.

Griffiths, D.V. (1985) *HARMONY – A program for predicting the elasto-plastic response of axisymmetric Bodies subjected to non-axisymmetric loading*, Report to Fugro-McClelland Engineers B.V., University of Manchester, UK.

Guo, W., Chu, J. and Kou, H. (2016) *Model tests of soil heave plug formation in suction caisson. Proc. Institution of Civil Engineers, Geotechnical Engineering*, Vol. 169, no. GE2, pp. 214–223.

Gupta, B.K. and Basu, D. (2016) *Design charts for laterally loaded rigid piles in elastic soil. Proc. ASCE Geo-Chicago 2016 GSP 270*, Chicago, Illinois, USA, pp. 393–406.

Hansteen, O.E., Jostad, J.P. and Tjelta, T.I. (2003) *Observed platform response to a "monster" wave. Proc. 6th Int. Symposium on Field Measurements in Geomechanics*, Swets and Zeitlinger, Lisse, The Netherlands.

Heins, E. and Grabe, J. (2017) Class-A-prediction of lateral pile deformation with respect to vibratory and impact pile driving. *Computers and Geotechnics*, Vol. 86, pp. 108–119.

Helfrich, S.C., Brazill, R.L. and Richards, A.F. (1976) *Pullout characteristics of a suction anchor in sand. Proc. of Offshore Technology Conference*, Houston, Texas, USA, Paper OTC-2469-MS.

Hernandez-Martinez, F.G., Rahim, A., Strandvik, S., Jostad, H.P. and Andersen, K.H. (2009) *Consolidation settlement of skirted foundations for subsea structures in soft clay. Proc. 17th International Conference on Soil Mechanics and Geotechnical Engineering*, pp. 1167–1172.

Hernandez-Martinez, F.G., Saue, M., Schroder, K. and Jostad, H.P. (2015) *Trenching effects on holding capacity for In-service suction anchors in high plasticity clays.*

Proc. 20th Offshore Symposium, Texas Section of the Society of Naval Architects and Marine Engineers, Houston, Texas, USA.

Hogervorst, J.R. (1980) *Field trials with large diameter suction piles. Proc. of Offshore Technology Conference*, Houston, Texas, USA, Paper OTC-3817-MS.

Holeyman, A., Peralta, P. and Charue, N. (2015) *Boulder-soil-pile dynamic interaction. Proc. 3rd International Symposium on Frontiers in Offshore Geotechnics*, Taylor & Francis, Oslo, Norway, pp. 563–568.

Hölscher, P., van Tol, A.F. and Huy, N.Q. (2009) *Influence of rate effect and pore water pressure during rapid load test of piles in sand. Proc. Rapid Load Testing on Piles*, van Tol, F. and Hölscher, P. (Eds.), CRC Press, pp. 59–72.

Holtz, R.D. (1991) Stress distribution and settlement of shallow foundations. In: *Foundation Engineering Handbook*, Winterkorn, H.F. and Fang, H.Y. (Eds.), 2nd edn, Van Nostrand Reinhold Company, New York, pp. 166–222.

Horvath, R.G. (1978) *Field load test data on concrete to rock bond strength*, University of Toronto, Canada, Publication No. 78-07. 23.

Horvath, R.G. and Kenney, T.C. (1979) *Shaft resistance of rock-socketed drilled piers. Symposium on Deep Foundations, ASCE National Convention*, Atlanta, Georgia, USA, pp. 182–214.

Hossain, M., Lehane, B., Hu, Y. and Gao, Y. (2012) Soil flow mechanisms around and between stiffeners of caissons during installation in clay. *Canadian Geotechnical Journal*, Vol. 49, no. 4, pp. 442–459.

Houlsby, G.T. and Byrne, B.W. (2005) *Design procedures for installation of suction caissons in sand. Proc. Institution of Civil Engineers, Geotechnical Engineering*, Vol. 158, no. GE3, Paper 13818, pp. 135–144.

Houlsby, G.T., Kelly, R.B. and Byrne, B.W. (2005) *The tensile capacity of suction caissons under rapid loading. Proc. 1st International Symposium on Frontiers in Offshore Geotechnics*, Taylor & Francis, Perth, Australia, pp. 405–410.

Houlsby, G.T., Kelly, R.B., Huxtable, J. and Byrne, B.W. (2006) Field trials of suction caissons in sand for offshore wind turbine. *Géotechnique*, Vol. 56, no. 1, pp. 3–10.

Houlsby, G.T. and Martin, C.T. (2003) Undrained bearing capacity factors for conical footings on clay. *Géotechnique*, Vol. 53, no. 5, pp. 513–520.

Houlston, P., Manceau, S., Benson, A. Alobaidi, I., Evans, R., Pacheco Andrade, M. and Meincke, J.H. (2017) *Laboratory testing based design method for assessing cyclic response of driven piles in chalk. Proc. of the 8th Int. Conf. Offshore Site Investigation and Geotechnics*. OSIG 2017, Society of Underwater Technology, pp. 449–456.

Huy, N.Q., van Tol, A.F. and Hölscher, P. (2009) *Rapid model pile load tests in the geotechnical centrifuge. Proc. Rapid Load Testing on Piles*, van Tol, F. and Hölscher, P. (Eds.), CRC Press, pp. 103–127.

Ibsen, C.B. and Thilsted, C.L. (2010) *Numerical study of piping limits for suction installation of offshore skirted foundations and anchors in layered sand. Proc. 2nd International Symposium on Frontiers in Offshore Geotechnics*, Perth, Australia, pp. 421–426.

ICE (1992) Bothkennar soft clay test site: characterisation and lessons learned, 8th Géotechnique Symposium in Print. *Géotechnique*, Vol. 42, no. 2, pp. 163–375.

IHC (2010) *IHC Hydrohammer. Pile Driving Equipment*, IHC Merwede, The Netherlands, 13 pp.

Irvine, J., Terente, V., Lee, L.T. and Comrie, R. (2015) *Driven pile design in weak rock. Proc. 3rd International Symposium on Frontiers in Offshore Geotechnics*, Taylor & Francis, Oslo, Norway, pp. 569–574.

ISO (2006) *Petroleum and natural gas industries – fixed concrete offshore structures*, International Organization for Standardization, ISO 19903-1:2019.

ISO (2007) *Petroleum and natural gas industries – fixed steel offshore structures*, International Organization for Standardization, ISO 19902:2007.

ISO (2011) *Petroleum and natural gas industries – specific requirements for offshore structures – part 7: stationkeeping systems for floating offshore structures and mobile offshore units, International Organization for Standardization*, ISO/DIS 19901-7.

ISO (2014a) *Petroleum and natural gas industries – specific requirements for offshore structures – part 7: stationkeeping systems for floating offshore structures and mobile offshore units*, ISO 19901-7:2013.

ISO (2014b) *Petroleum and natural gas industries – specific requirements for offshore structures – part 8: marine soil investigations International Organization for Standardization*, ISO 19901-8:2014.

ISO (2016a) *Petroleum and natural gas industries – specific requirements for offshore structures – part 4: geotechnical and foundation design considerations*, International Organization for Standardization, ISO 19901-4:2016.

ISO (2016b) *Petroleum and natural gas industries – site specific assessment of mobile offshore units - part 1: jack-ups*, International Organization for Standardization, ISO 19905-1:2016(E).

Janbu, N. (1985) Soil models in offshore engineering. *Géotechnique*, Vol. 35, no. 3, pp. 241–281.

Jardine, R., Chow, F.C., Overy, R.F. and Standing, J.R. (2005) *ICP design methods for driven piles in sands and clays*, Thomas Telford Ltd., London.

Jardine, R.J., Buckley, R.M., Byrne, B.W., Kontoe, S., McAdam, R.A., Andolfsson, T., Liu, T., Schranz, F. and Vinck, K. (2019) Improving the design of piles driven in chalk through the ALPACA research project. *Revue Française de Géotechnique*, Vol. 158, no. 2. doi:10.1051/geotech/2019008.

Jardine, R.J., Standing, J.R. and Chow, F.C. (2006) Some observations of the effects of time on the capacity of piles driven in sand. *Géotechnique*, Vol. 56, no. 4, pp. 227–244.

Jeanjean, P. (2006) *Set-up characteristics of suction anchors for soft Gulf of Mexico clays: experience from field installation and retrieval. Proc. Offshore Technology Conference*, Houston, Texas, USA, Paper OTC-18005-MS.

Jeanjean, P., Znidarcic, D., Phillips, R., Ko, H-Y., Pfister, S., Cinicioglu, O. and Schroeder, K. (2006) *Centrifuge testing on suction anchors: Double-wall, over-consolidated clay, and layered soil profile. Proc. Offshore Technology Conference*, Houston, Texas, USA, Paper OTC-18007-MS.

Jostad, H.P. (1997) *BIFURC - Version 3 Undrained Capacity Analyses of Clay*, Technical Report, Norwegian Geotechnical Institute.

Kallehave, D., LeBlanc, C., Liingaard, M.A. (2012) *Modification of the API P-Y formulation of initial stiffness of sand, Integrated Geotechnologies - Present and Future. Proc. of the 7th Int. Conf. Offshore Site Investigation and Geotechnics*. OSIG 2017, Society of Underwater Technology, London, pp. 465–472.

Karlsrud, K., Jensen, T.G., Lied, E.K.W., Nowacki, F. and Simonsen, A.S. (2014) *Significant ageing effects for axially loaded piles in sand and clay verified by new field load tests. Proc. Offshore Technology Conference*, Houston, Texas, USA, Paper OTC-25197-MS.

Kay, S. (2013) *Torpedo piles – VH capacity in clay using resistance envelope equations. Proc. 32nd International Conference on Ocean, Offshore and Arctic Engineering*, OMAE2013, 9–14 June, Nantes, France.

Kay, S. (2015) *CAISSON_VHM: a program for caisson capacity under VHM Load in undrained soils. Proc. 3rd International Symposium on Frontiers in Offshore Geotechnics*, Taylor & Francis, Oslo, Norway, pp. 277–282.

Kay, S. (2018) *User's Manual CAISSON_VHM Version 00.12, Computer Program for Caisson Foundation Resistance in Undrained Soils during VHM Load*, S.Kay Consultant Report SKA/MAN/003, March, 850 pp.

Kay, S. and Palix, E. (2010) *Caisson capacity in clay: VHM resistance envelope Part 2: VHM envelope equation and design procedures. Proc. 2nd International Symposium on Frontiers in Offshore Geotechnics*, Perth, Australia.

Kay, S. and Palix, E. (2011) *Caisson capacity in clay: VHM Resistance Envelope – Part 3: extension to shallow foundations. Proc. 30th International Conference on Ocean, Offshore and Arctic Engineering*, OMAE2011, 19–24 June, Rotterdam, the Netherlands.

Kay, S. and Van Woensel, F. (2002) *Cone pressuremeter tests in clay. Proc. 5th European Conference Numerical Methods in Geotechnical Engineering*, Paris, 4–6 September 2002, Laboratoire Central des Ponts et Chaussées (LCPC) and Presses de l'École Nationale des Ponts et Chaussées, Paris, pp. 99–105.

Keaveny, J.V., Hansen, S.B., Madshus, C. and Dyvik, R. (1994) *Horizontal capacity of large-scale model anchors. Proc. 13th Int. Conf. Soil Mechanics and Foundation Engineering*, New Delhi, India, pp. 677–680.

Kellezi, L., Kudsk, G. and Hansen, P.B. (2005) *FE modelling of spudcan-pipeline interaction. Proc. 1st International Symposium on Frontiers in Offshore Geotechnics*, Taylor & Francis, Perth, Australia, pp. 551–557.

Kelly, R.B., Byrne, B.W., Houlsby, G.T. and Martin, C.M. (2003) *Pressure chamber testing of model caisson foundations in sand. Proc. 2nd British Geotechnical Association International Conference on Foundations*, 2–5 September, University of Dundee, Scotland, UK, IHS BRE Press, pp. 421–432.

Kelly, R.B., Byrne, B.W., Houlsby, G.T. and Martin, C.M. (2004) *Tensile loading of model caisson foundations for structures on sand. Proc. 14th International Offshore and Polar Engineering Conference*, Toulon, Vol. 2, pp. 638–641.

Kelly, R.B., Houlsby, G.T. and Byrne, B.W. (2006a) A comparison of field and laboratory tests of caisson foundations in sand and clay. *Géotechnique*, Vol. 56, no. 9, pp. 617–626.

Kelly, R.B., Houlsby, G.T. and Byrne, B.W. (2006b) Transient vertical loading of model suction caissons in a pressure chamber. *Géotechnique*, Vol. 56, no. 10, pp. 665–675.

Kennedy, J., Oliphant, J., Maconochie, A., Stuyts, B. and Cathie, D. (2013) *CAISSON: a suction pile design tool. Proc. 32nd International Conference on Ocean, Offshore and Arctic Engineering*, OMAE2013, 9–14 June, Nantes, France.

Kirsch, F., Richter, T., Coronel, M. (2014) Geotechnische Aspekte bei der Gründungsbemessung von Offshore-Windenergieanlagen auf Monopfählen mit sehr großen Durchmessern, Stahlbau Spez. 2014 – Erneuerbare Energien 83.

Kolk, H.J., Kay, S., Kirstein, A. and Troestler, H. (2001) *North Nemba flare bucket foundations. Proc. Offshore Technology Conference*, Houston, Texas, USA, Paper OTC-13057-MS.

Kolk, H.J. and Wegerif, J. (2005) *Offshore site investigations: new frontiers. Proc. 1st International Symposium on Frontiers in Offshore Geotechnics*, Taylor & Francis, Perth, Australia, pp. 145–161.

Kuhlemeyer, R. (1979) Static and dynamic laterally loaded floating piles. *ASCE Journal of Geotechnical and Geoenvironmental Engineering*, Vol. 105, 289–304.

Lambe, T.W. (1973) Predictions in soil engineering. *Géotechnique*, Vol. 23, no. 2, pp. 151–201.

Langford, T.E., Solhjell, E., Hampson, K. and Hondebrink, L. (2012) *Geotechnical design and installation of suction anchors for the Skarv FPSO, Offshore Norway. Proc. 7th Int. Conf. Offshore Site Investigation and Geotechnics*. OSIG 2012, Society of Underwater Technology, London, pp. 613–620.

LeBlanc, C., Houlsby, G.T., Byrne, B.W. (2010) Response of stiff piles in sand to long-term cyclic lateral loading. *Géotechnique*, Vol. 60, no. 2, pp. 79–90.

Lee, Y.C., Audibert, J.M.E. and Tjok, K.-M. (2005) *Lessons learned from several suction caisson installation projects in clay. Proc. 1st International Symposium on Frontiers in Offshore Geotechnics*, Taylor & Francis, Perth, Australia, pp. 235–241.

Lehane, B.M., Elkhatib, S. and Terzaghi, S. (2014) Extraction of suction caissons in sand, *Géotechnique*, Vol. 64, no. 9, pp. 735–739.

Lehane, B.M., Schneider, J.A. and Xu, X. (2005) *A review of design methods for offshore driven piles in siliceous sand*, University of Western Australia, Perth, UWA Report, GEO:05358.

Lin, S.-S. and Liao, J.-C. (1999) Permanent strains of piles in sand due to cyclic lateral loads. *ASCE Journal of Geotechnical and Geoenvironmental Engineering*, Vol. 125, No. 9, pp. 798–802.

Lovera, A. (2019) *Cyclic lateral design for offshore monopiles in weak rocks*. PhD Thesis, Université Paris-Est, Environmental Engineering, France. NNT: 2019PESC1016.

Lovera, A., Ghabezloo, S., Sulem, J., Randolph, M.F. and Palix, E. (2020) *Extension of the p-y curves framework for cyclic loading of offshore wind turbines monopiles. Proc. 4th International Symposium on Frontiers in Offshore Geotechnics*, Austin, Texas, USA.

Lunne, T., Robertson, P.K. and Powell, J.J.M. (1997) *Cone penetration testing in geotechnical practice*, Blackie Academic & Professional, London.

Lunne, T. and St John, H.D. (1979) *The use of cone penetrometer tests to compute penetration resistance of steel skirts underneath North Sea gravity platforms. Proc. 7th European Conference on Soil Mechanics and Foundation Engineering*, Brighton, England, September, Vol. 2, British Geotechnical Society, London, pp. 233–238.

Lunne, T.A., Myrovll, F. and Kleven, A. (1982) *Observed settlements of five North Sea gravity platforms*, pp. 25, NGI Publication 139. ISBN: 0078-1193. A Norwegian Geotechnical Institute (NGI) publication, Norway.

Madsen, S., Andersen, L.V. and Ibsen, L.B. (2012) *Instability during installation of foundations for offshore structures. Proc. 16th Nordic Geotechnical Meeting*, Copenhagen, Denmark, Vol. 2, Dansk Geoteknisk Forening, pp. 499–505.

Mana, D.S.K., Gourvenec, S. and Randolph, M.F. (2013) A novel technique to mitigate the effect of gapping on the uplift capacity of offshore shallow foundations. *Géotechnique*, Vol. 63, no. 14, pp. 1245–1252.

Martinelli, M., Alderlieste, E.A., Galavi, V. and Luger, H.J. (2020) *Numerical simulation of the installation of suction bucket foundations using MPM. Proc. 4th International Symposium on Frontiers in Offshore Geotechnics*, Austin, Texas, USA.

McManus, K.J. and Davis, R.O. (1997) Dilation-induced pore fluid cavitation in sands. *Géotechnique*, Vol. 47, no. 1, pp. 173–177.

McNamara, A.P. (2000) *Behaviour of soil around the internal stiffeners of suction caissons and the effect on installation and pullout resistance*, Geomechanics Group Report No. G1512, University of Western Australia, Australia, June.

Meigh, A.C. and Wolski, W. (1979) *Design parameters for weak rock. Proc. 7th European Conference on Soil Mechanics and Foundation Engineering*, British Geotechnical Society, London, UK.

Metrikine, A.V., Tsouvalas, A, Segeren, M.L.A., Elkadi, A.S.K., Tehrani, F.S., Gómez, S.S., Atkinson, R., Pisanò, F., Kementzetzidis, E., Tsetas, A., Molenkamp, T., van Beek, K. and de Vries, P. (2020) *GDP: a new technology for gentle driving of (mono)piles. Proc. 4th International Symposium on Frontiers in Offshore Geotechnics*, article #3483; publication #1069 (IC-ISFOG21), 8–11 November 2021, Austin, Texas, USA, ASCE Geo-Institute and DFI (Proc. published in August 2020).

Mirza, U.A., Sweeney, M. and Dean, A.R. (1988) *Potential effects of jack-up spudcan penetration on jacket piles. Proc. of Offshore Technology Conference*, Houston, Texas, USA, OTC-5762-MS.

Muir Wood, A., Mackenzie, B., Burbury, D., Rattley, M., Clayton, C.R.I., Mygind, M., Wessel Andersen, K., LeBlanc Thilsted, C. and Alberg Liingaard, M. (2015) *Design of large diameter monopiles in chalk at Westermost Rough offshore wind farm. Proc. 3rd International Symposium on Frontiers in Offshore Geotechnics*, Taylor & Francis, Oslo, Norway, pp. 723–728.

Murff, J.D. (1975) Response of axially loaded piles. *Journal of the Geotechnical Engineering Division, Proceedings of the American Society of Civil Engineers*, Vol. 101, no. GT3, pp. 356–360.

Murff, J.D. (1994) *Limit analysis of multi-footing foundation systems. Proc. Int. Conf. on Computer Methods and Advances in Geomechanics*, Morgantown, 1, pp. 223–244.

Murff, J.D. and Hamilton, J.M. (1993) P-ultimate for undrained analysis of laterally loaded pile'. *ASCE Journal of Geotnical Engineering*, Vol. 119, no. 1, pp. 91–107.

Negro, V., López-Gutiérrez, J-S., Dolores Esteban, M., Alberdi, P., Imaz, M. and Serraclara, J-M. (2017) Monopiles in offshore wind: preliminary estimate of main dimensions. *Ocean Engineering*, Vol. 133, pp. 253–261.

Neubecker, S.R. and Randolph, M.F. (1995) Profile and frictional capacity of embedded anchor chain. *ASCE Journal of Geotnical Engineering*, Vol. 121, no. 11, pp. 787–803.

Newlin, J.A. (2003a) *Suction anchor piles for the Na Kika FDS mooring system. Part 1: site characterization and design. Proc. International Symposium on Deepwater Mooring Systems: Concepts, Design, Analysis and Materials*, American Society of Civil Engineers, Houston, Texas, USA, pp. 28–54.

Newlin, J.A. (2003b) *Suction anchor piles for the Na Kika FDS mooring system. Part 2: site installation performance. Proc. International Symposium on Deepwater*

Mooring Systems: Concepts, Design, Analysis and Materials, American Society of Civil Engineers, Houston, Texas, USA, pp. 55–75.

NGI (1997) *Skirted foundations and anchors in clay – state of the art report*, Norwegian Geotechnical Institute Report 524071-1, Norway, 15 December, p. 158.

Nguyen, T.C., van Lottum, H., Hölscher, P. and van Tol, A.F. (2012) *Centrifuge modeling of rapid load tests with open-ended piles. Proc. 9th Int. Conf. on Testing and Design Methods for Deep Foundations*, Japan, pp. 265–272.

Nichols, G. (2009) *Sedimentology and stratigraphy*, 2nd edn, Wiley-Blackwell, Canada, 432 pp.

Norwegian Geotechnical Institute (1997) *Skirted foundations and anchors in clays - State-of-the-art*, NGI Report No. 52407-1, December.

Norwegian Geotechnical Institute (2004) *Windows program HVMCap. Version 3.0: theory, user manual and certification.* Norwegian Geotechnical Institute Report 524096-7, Rev. 2.

Novello, E.A. (1999) *From static to cyclic p-y data in calcareous sediments. Proc. 2nd International Conference on Engineering for Calcareous Sediments*, 15–18 March, Routledge, Perth, pp. 17–27.

NPD (1992) *Guidelines concerning loads and load effects to regulations concerning loadbearing structures in the petroleum activities.* Issued by the Norwegian Directorate, 7 February.

O'Loughlin, C.D., Randolph, M.F. and Richardson, M. (2004) *Experimental and theoretical studies of deep penetrating anchors. Proc. Offshore Technology Conference*, Houston, Texas, USA, Paper OTC-16841-MS.

O'Riordan, N.J. (1991) Effects of cyclic loading on the long-term settlements of structures, Chapter 9. In: *Cyclic loading of soils: from theory to design*, O'Reilly, M.P. and Brown, S.F. (Eds.), Blackie & Son Ltd, UK, pp. 411–433.

Osman, A.S. and Randolph, M.F. (2012) Analytical solution for the consolidation around a laterally loaded pile. *ASCE International Journal of Geomechanics*, Vol. 12, no. 3, pp. 199–208.

OTRC (2008) *FALL16 PROGRAM. Foundation analysis by the limit load method -- 16th in a series of limit analysis programs. Suction caisson analysis under inclined load*, Excel document FALL16Rev3Mar2008_master.xls.

Paikowsky, S.G. (1989) *A static evaluation of soil plug behavior with application to the pile plugging problem*, PhD. Thesis, Massachusetts Institute of Technology, USA.

Palix, E., Chan, N., Yangrui, Z. and Haijing, W. (2013) *Liwan 3-1: how deepwater sediments from South China Sea compare with Gulf of. Guinea sediments. Proc. Offshore Technology Conference*, Houston, Texas, USA, Paper OTC-24010-MS.

Palix, E. and Lovera, A. (2020) *Field testing for monopile to be installed in weak carbonated rock. Proc. 4th International Symposium on Frontiers in Offshore Geotechnics*, Austin, Texas, USA.

Palix, E., Willems, T. and Kay, S. (2010) *Caisson capacity in clay: VHM resistance envelope – Part 1: 3D FEM numerical study. Proc. 2nd International Symposium on Frontiers in Offshore Geotechnics*, Perth, Australia, pp. 753–758.

Panagoulias, S. (2016) *Critical pressure during installation of suction caissons in sand*, MSc. Thesis, Delft University of Technology, The Netherlands.

Panagoulias, S., van Dijk, B.F.J., Drummen, T., Askarinejad, A. and Pisanò, F. (2017) *Critical suction pressures during installation of suction caissons in sand. Proc. of the 8th Int. Conf. Offshore Site Investigation and Geotechnics.* OSIG 2017, Society of Underwater Technology, London, pp. 570–577.

PanGeo (2010) *Mitigating risk for offshore installations*, presentation to *The Hydographic Society, UK*, 21 October 2010, 54 pp. https://www.ths.org.uk/event_details.asp?v0=234.

Peck, R.B. (1969) Advantages and limitations of the observational method in applied soil mechanics. *Géotechnique*, Vol. 19, no. 2, pp. 171–187.

Pells, P.J.N., Rowe, R.K. and Turner, R.M. (1980) *An experimental investigation into side shear for socketed piles in sandstone. Proc. International Conference on Structural Foundations on Rock*, Sydney, 7–9 May, Balkema, Vol. 1, pp. 291–302.

Pile Dynamics, Inc. (2010) *GRLWEAP, wave equation analysis of pile driving*, Computer Program Package.

Poulos, H.G. and Davis, E.H. (1980) *Pile foundation analysis and design*, John Wiley and Sons, New York, Series in Geotechnical Engineering, 410 pp.

Poulos, H.G. and Davis, E.H. (1991) *Elastic solutions for soil and rock mechanics*, Series in Soil Engineering, Originally published 1974 by John Wiley & Sons (ISBN 10: 0471695653 ISBN 13: 9780471695653), republished in 1991 by Centre for Geotechnical Research, Sydney University. Available online http://www.usucger.org/PandD/PandD.htm.

Poulos, H.G. and Randolph, M.F. (1983) Pile group analysis: a study of two methods. *ASCE, Journal of Geotechnical Engineering*, Vol. 109, no. 3, March, pp. 355–372.

Puech, A., Becue, J.P. and Colliat, J.-L. (1988) *Advances in the design of piles driven into non-cemented to weakly cemented carbonate formations. Proc. 1st International Conference on Calcareous Sediments*, 15–18 March 1988, Perth, A.A. Balkema, Rotterdam, pp. 305–312.

Puech, A., Colliat, J-L., Nauroy, J.F. and Meunier, J. (2005) *Some geotechnical specificities of Gulf of Guinea deepwater sediments. Proc. 1st International Symposium on Frontiers in Offshore Geotechnics*, Taylor & Francis, Perth, Australia, pp. 1047–1053.

Puech, A., Poulet, D. and Boisard, P. (1990) *A procedure to evaluate pile drivability in the difficult soil conditions of the southern part of the Gulf of Guinea. Proc. Offshore Technology Conference*, Houston, Texas, USA, Paper OTC-6237-MS.

Puech, A. and Quiterio-Mendoza, B. (2019) Caractérisation des massifs rocheux pour le dimensionnement de pieux forés en mer (Characterization of rock masses for designing drilled and grouted offshore pile foundations). *Revue Française de Géotechnique*, Vol. 158, no. 5. doi:10.1051/geotech/2019011.

Randolph, M.F. (1981) Piles subjected to torsion. *Journal of the Geotechnical Engineering Division, ASCE*, Vol. 107, no. GT8, pp. 1095–1111.

Randolph, M.F. (2018) Potential damage to steel pipe piles during installation. *International Press-in Association (IPA)*, Vol. 3, no. 1, March 2018, pp. 8, Special Contribution. https://www.press-in.org/en/page/specialcontributions.

Randolph, M.F. and Gourvenec, S. (2011) *Offshore geotechnical engineering*, Spon Press, London.

Randolph, M.F. and Houlsby, G.T. (1984) The limiting pressure on a circular pile loaded laterally in cohesive soil. *Géotechnique*, Vol. 34, no. 4, pp. 613–623.

Randolph, M.F. and House, A.R. (2002) *Analysis of suction caisson capacity in clay. Proc. Offshore Technology Conference*, Houston, Texas, USA, Paper OTC-14236-MS.

Randolph, M.F., Martin, C.M. and Hu, Y. (2000) Limiting resistance of a spherical penetrometer in cohesive material. *Géotechnique*, Vol. 50, no. 5, pp. 573–582.

Randolph, M.F., O'Neill, M.P. and Stewart, D.P. (1998) *Performance of suction anchors in fine-grained calcareous soils. Proc. Offshore Technology Conference*, Houston, Texas, USA, Paper OTC-8831-MS.

Randolph, M.F. and Wroth, C.P. (1978) Analysis of deformation of vertically loaded piles. *Journal of the Geotechnical Engineering Division, American Society of Civil Engineers*, Vol. 104, no. GT12, pp. 1465–1488.

Reese, L.C. (1997) Analysis of laterally loaded piles in weak rock. *ASCE Journal of Geotechnical and Geoenvironmental Engineering*, Vol. 123, no. 11, pp. 1010–1017.

Reese, L.C., Cox, W.R. and Koop, F.D. (1974) *Analysis of laterally loaded piles in sand. Proc. Offshore Technology Conference*, Houston, Texas, USA, Paper OTC-2080-MS.

Romp, R.H. (2013) *Installation-effects of suction caissons in non-standard soil conditions*, MSc. Thesis, Delft University of Technology, The Netherlands.

Rosenberg, P. and Journeaux, N.L. (1976) Friction and end bearing tests on bedrock for high capacity socket design. *Canadian Geotechnical Journal*, Vol. 13, pp. 324–333.

Rowe, R.K. and Armitage, H.H. (1987) A design method for drilled piers in soft rock. *Canadian Geotechnical Journal*, Vol. 24. pp. 126–142.

Salgado, R., Lyamin, A.V., Sloan, S.W. and Yu, H.S. (2004) Two and three-dimensional bearing capacity of foundations in clay. *Géotechnique*, Vol. 54, no. 5, pp. 297–306.

Sassi, K., Kuo, M.Y.-H., Versteele, H., Cathie, D. and Zehzouh, S. (2017) *Insights into the mechanisms of anchor chain trench formation. Proc. 8th Int. Conf. Offshore Site Investigation and Geotechnics*, OSIG 2017, Society of Underwater Technology, London, pp. 963–970.

Sassi, K., Zehzouh, S., Blanc, M., Thorel, L., Cathie, D., Puech, A. and Colliat-Dangus, J-L. (2018) *Effect of seabed trenching on the holding capacity of suction anchors in soft deepwater Gulf of Guinea clays – Centrifuge testing and numerical modelling. Proc. Offshore Technology Conference*, Houston, Texas, USA, Paper OTC-28756-MS.

Saviano, A. (2015) *Effects of misalignment on the undrained HV capacity of suction anchors in clay*, MSc. Thesis, TU Delft, The Netherlands.

Saviano, A. and Pisanò, F. (2017) Effects of misalignment on the undrained HV capacity of suction anchors in clay. *Ocean Engineering*, Vol. 133, pp. 89–106.

Schneider, J.A., White, D.J. and Lehane, B.M. (2007) *Shaft friction of piles in siliceous, calcareous and micaceous sands. Proc. 6th Int. Conf. Offshore Site Investigation and Geotechnics*. OSIG 2007, Society of Underwater Technology, London, pp. 367–382.

Seibold, S. and Berger, W.H. (2010) *The seafloor: an introduction to marine geology*, 3rd edn, Springer-Verlag, Germany, 272 pp.

Seidel, J.P. and Collingwood, B. (2001) A new socket roughness factor for prediction of rock socket shaft resistance. *Canadian Geotechnical Journal*, Vol. 38, no. 1, pp. 138–153.

Senders, M. (2005) *Tripods with suction caissons as foundations for offshore wind turbines on sand. Proc. 1st International Symposium on Frontiers in Offshore Geotechnics*, Taylor & Francis, Perth, Australia, pp. 397–405.

Senders, M. (2008) *Suction caissons in sand as tripod foundations for offshore wind turbines*, PhD thesis, School of Civil and Resource Engineering, UWA, Australia.

Senders, M. and Kay, S. (2002) *Geotechnical suction pile anchor design in deep water soft clays. Proc. 7th Annual Conference, Deepwater Risers, Moorings and Anchorings*, IBC, London, UK, 50 pp.

Senders, M. and Randolph, M.F. (2009) CPT-based method for the installation of suction caissons in sand. *ASCE Journal of Geotechnical and Geoenvironmental Engineering*, Vol. 135, no. 1, January, pp. 14–25.

Senders, M., Randolph, M.F. and Gaudin, C. (2007) *Theory for the installation of suction caissons in sand overlain by clay. Proc. 6th Int. Conf. Offshore Site Investigation and Geotechnics*. OSIG 2007, Society of Underwater Technology, London, pp. 429–438.

Senpere, D. and Auvergne, G.A. (1982) *Suction anchor piles – a proven alternative to driving or drilling. Proc. Offshore Technology Conference*, Houston, Texas, USA, Paper OTC-4206-MS.

Shonberg, A., Harte, I., Aghakouchak, A., Brown, C.S.D, Andrade, M.P. and Liingaard, M.A. (2017) *Suction bucket jackets for offshore wind turbines: applications from in situ observations. Proc. TC 209 Workshop - 19th ICSMGE*, 20 September 2017, Seoul, pp. 65–77.

Siciliano, R.J., Hamilton, J.M., Murff, J.D. and Phillips, R. (1990) *Effects of jackup spudcans on piles. Proc. Offshore Technology Conference*, Houston, Texas, USA, Paper OTC-6467-MS.

Simons, N., Menzies, B. and Matthews, M. (2002) *A short course in geotechnical site investigation*, Thomas Telford Publishing, London, ISBN 9780727729484.

Skempton, A.W. (1951) *The bearing capacity of clays. Proc. Building Research Congress*, Vol. 1, pp. 180–189.

Smith, E.A.L. (1960) Pile-driving analysis by the wave equation. *Journal of the Soil Mechanics and Foundations Division ASCE*, Vol. 86, no. SM4, pp. 35–61.

Smith, I.M. and Chow, Y.K. (1982) *Three-dimensional analysis of pile drivability. Proc. First Int. Conf. Num. Methods in Offshore Piling*, ICE, London, UK, 20 pp.

Smith, I.M. and Griffiths, D.V. (1998) *Programming the finite element method*, 3 rd edn, Wiley, USA, 534 pp.

Smith, I.M., Hicks, M.A., Kay, S. and Cuckson, J. (1988) *Undrained and partially drained behaviour of end bearing piles and bells founded in untreated calcarenite. Proc. 1st International Conference on Calcareous Sediments*, 15–18 March 1988, Perth, Vol. 2, A.A. Balkema, Rotterdam, pp. 663–679.

SOLCYP (2017) *Design of piles under cyclic loading – SOLCYP recommendations*, Puech, A. and Garnier, J. (Ed.), ISTE – WILEY, France.

Sorensen, S.P.H., Ibsen, L.B., Augustesen, A.H. (2010) *Effects of diameter on initial stiffness of p-y curves for large-diameter piles in sand, in: numerical Methods in Geotechnical Engineering*. CRC Press, pp. 907–912.

Sørlie, E. (2013) *Bearing capacity failure envelope of suction caissons subjected to combined loading*, MSc. Thesis, Norwegian University of Science and Technology, Trondheim.

Stavropolou, E., Dano, C. and Boulon, M. (2019) *Experimental investigation of the mechanical behaviour of soft carbonate rock/grout interfaces for the design of offshorewind turbines. Proc. E3S Web of Conferences 7th Int. Symp. Deformation Characteristics of Geomaterials*, Vol. 92, IS-Glasgow. doi:10.1051/e3sconf/20199213006.

Stevens, J.B., Audibert, J.M. (1979) *Re-examination of PY curve formulations. Proc. Offshore Technology Conference*, Houston, Texas, USA, Paper OTC-3402-MS.

Stevens, R.S., Wiltsie, E.A. and Turton, T.H. (1982) *Evaluating pile drivability for hard clay, very dense sand and rokc. Proc. Offshore Technology Conference*, Houston, Texas, USA, Paper OTC-4205-MS.

Stewart, D. (2005) *Influence of jack-up operation adjacent to a piled structure. Proc. 1st International Symposium on Frontiers in Offshore Geotechnics*, Taylor & Francis, Perth, Australia pp. 543–550.

Sturm, H. (2017) *Design aspects of suction caissons for offshore wind turbine foundations. Proc. TC 209 Workshop – 19th ICSMGE*, 20 September 2017, Seoul, pp. 45–63.

Sturm, H., Nadim, F. and Page, A. (2015) *A safety concept for penetration analyses of suction caissons in sand. Proc. 3rd International Symposium on Frontiers in Offshore Geotechnics*, Taylor & Francis, Oslo, Norway, pp. 1393–1398.

Supachawarote, C. (2006) *Inclined load capacity of suction caissons in clay*, PhD thesis. University of Western Australia, Australia.

Supachawarote, C., Randolph, M.F. and Gourvenec, S. (2004) *Inclined pull-out capacity of suction caissons. Proc. 14th International Offshore and Polar Engineering Conference ISOPE*, 23–28 May 2004, Toulon, France, International Society of Offshore and Polar Engineers (ISOPE), Cupertino, pp. 500–506.

Suroor, H. and Hossain, J. (2015) *Effect of torsion on suction piles for subsea and mooring applications. Proc. 3rd International Symposium on Frontiers in Offshore Geotechnics*, Taylor & Francis, Oslo, Norway, pp. 325–330.

Svanø, G., Eiksund, G., Kavli, A., Langø, H., Karunakaran, D. and Tjelta, T.I. (1997) Soil-structure interaction of the Draupner E bucket foundation during storm conditions. *BOSS*, Vol. 97, pp. 163–176.

Taiebat, H.A. and Carter, J.P. (2000) Numerical studies of the bearing capacity of shallow foundations on cohesive soil subjected to combined loading. *Géotechnique*, Vol. 50, no. 4, pp. 409–418.

Taiebat, H.A. and Carter, J.P. (2002) *A failure surface for the bearing capacity of circular footings on faturated clays. Proc. 8th Int. Symp. Numerical Models in Geomechanics (NUMOG)*, 10–12 April 2002, Rome, Italy, pp. 457–462.

Taiebat, H.A. and Carter, J.P. (2005) *A failure surface for caisson foundations in undrained soils. Proc. 1st International Symposium on Frontiers in Offshore Geotechnics*, Taylor & Francis, Perth, Australia, pp. 289–295.

Taiebat, H.A. and Carter, J.P. (2010) A failure surface for circular footings on cohesive soils. *Géotechnique*, Vol. 60, no. 4, pp. 265–273.

Terente, V., Torres, I., Irvine, J. and Jaeck, C. (2017) *Driven pile design methods in weak rock. Proc. 8th Int. Conf. Offshore Site Investigation and Geotechnics*. OSIG 2017, Society of Underwater Technology, London, pp. 652–657.

Terzaghi, K., Peck, R.B. and Mesri, G. (1996) *Soil Mechanics in Engineering Practice*, 3rd edn, Wiley, New York.

Tho, K.K., Chan, N. and Paisley, J. (2015) *Comparison of coupled and decoupled approaches to spudcan-pile interaction. Proc. 3rd International Symposium on Frontiers in Offshore Geotechnics*, Taylor & Francis, Oslo, Norway, pp. 1317–1322.

Tho, K.K., Leung, C.F., Chow, Y.K. and Swaddiwudhipong, S. (2013) *Methodologies for evaluation of spudcan-pile interaction. Proceedings of the 14th International Conference on Jack-Up Platform Design, Construction and Operation*, London, September.

Tjelta, T.I. (1995) *Geotechnical experience from the installation of the europipe jacket with bucket foundations. Proc. Offshore Technology Conference*, Houston, Texas, USA, Paper OTC-7795-MS.

Tjelta, T.I. (2015) *The suction foundation technology. Proc. 3rd International Symposium on Frontiers in Offshore Geotechnics*, Taylor & Francis, Oslo, Norway, pp. 85–93.

Tran, M.N. (2005) *Installation of suction caissons in dense sand and the influence of silt and cemented layers*, PhD. Thesis, Department of Civil Engineering, University of Sydney, Australia.

Tran, M.N., Randolph, M.F. and Airey, D.W. (2004) *Experimental study of suction installation of caissons in dense sand. Proc. 23rd International Conference on Offshore Mechanics and Arctic Engineering*, 20–25 June 2004, Vancouver, Canada, Paper OMAE2004-51076.

Tran, M.N., Randolph, M.F. and Airey, D.W. (2007) Installation of suction caissons in sand with silt layesr. *ASCE Journal of Geotechnical and Geoenvironmental Engineering*, Vol. 133, no. 10, October 1, pp. 1183–1191.

Van den Berg, P. (1994) *Analysis of soil penetration*, PhD. Thesis, TU Delft, The Netherlands.

Van Dijk, B.F.J. (2015) *Caisson capacity in undrained soil: failure envelopes with internal scooping. Proc. 3rd International Symposium on Frontiers in Offshore Geotechnics*, Taylor & Francis, Oslo, Norway, pp. 337–342.

Versteele, H., Jaeck, C., Silvano, R. and Bignold, D. (2015) *Sinkage analysis of A-shaped jack-up mudmats using the coupled Eulerian Lagrangian approach. Proc. 3rd International Symposium on Frontiers in Offshore Geotechnics*, Taylor & Francis, Oslo, Norway, pp. 1323–1328.

Versteele, H., Kuo, M.Y.-H., Cathie, D., Sassi, K. and Zehzouh, S. (2017) *Anchor chain trenching – numerical simulation of progressive erosion. Proc. 8th Int. Conf. Offshore Site Investigation and Geotechnics*, OSIG 2017, Society of Underwater Technology, London, pp. 971–977.

Vicente, V., López-Gutiérrez, J-S., Dolores Esteban, M., Alberdi, P., Imaz, M. and Serraclara, J-M. (2017) Monopiles in offshore wind: preliminary estimate of main dimensions. *Ocean Engineering*, Vol. 133, pp. 253–261.

Villalobos, F.A. (2007) *Installation of suction caissons in sand. Proceedings 6th Chilean Conf. on Geotechnics (Congreso Chileno de Geotecnia)*, Chile.

Vivatrat, V., Valent, P.J. and Ponterio, A.A. (1982) *The influence of chain friction on anchor pile design. Proc. Offshore Technology Conference*, Houston, Texas, USA, Paper OTC-4178-MS.

Vulpe, C., Gourvenec, S. & Power, M. (2014) A generalized failure envelope for undrained capacity of circular shallow foundations under general loading. *Géotechnique Letters*, Vol. 4, no. July–September, pp. 187–196.

Waltham, T. (2009) *Foundations of Engineering Geology*, 3rd edn, CRC Press, London. doi:10.1201/9781315273488.

Wang, M.C., Demars, K.R. and Nacci, V.A. (1977) Breakout capacity of model suction anchors in soil. *Canadian Geotechnical Journal*, Vol. 14, no. 2, pp. 246–257.

Wang, M.C., Demars, K.R. and Nacci, V.A. (1978) *Applications of suction anchors in offshore technology. Proc. Offshore Technology Conference*, Houston, Texas, USA, Paper OTC-3203-MS.

Wang, M.C., Nacci, V.A. and Demars, K.R. (1975) Behavior of underwater suction anchor in soil. *Ocean Engineering*, Vol. 3, no. 1, pp. 47–62.

Watson, P.G., Gaudin, C., Senders, M. and Randolph, M.F. (2006) *Installation of suction caissons in layered soil. Proc. 6th International Conference on Physical Modeling in Geotechnics*, Hong Kong, pp. 685–692.

Whitehouse, R.J.S., Harris, J.M., Sutherland, J. and Rees, J. (2011) The nature of scour development and scour protection at offshore windfarm foundations. *Marine Pollution Bulletin*, Vol. 62, no. 1, pp. 73–88.

Williams, A.F. and Pells, P.J.N. (1981) Side resistance rock sockets in sandstone, mudstone, and shale. *Canadian Geotechnical Journal*, Vol. 18, no. 4, pp. 502–513.

Wind Europe (2018) *Offshore wind in Europe - Key trends and statistics 2017*, February, 37 pp.

Wind Europe (2020) *Offshore wind in Europe - Key trends and statistics 2019*, https://windeurope.org/data-and-analysis/product/offshore-wind-in-europe-key-trends-and-statistics-2019/.

XG Geotools (2014) *SPCalc Manual*.

Xie, Y. (2009) *Centrifuge model study on spudcan-pile interaction*, PhD. Thesis, Department of Civil and Environmental Engineering, National University of Singapore, Singapore.

Xie, Y, Leung, C.F. and Chow, Y.K. (2017) Centrifuge modelling of spudcan-pile interaction in soft clay overlying sand. *Géotechnique*, Vol. 67, no. 1, pp. 69–77.

Zienkiewicz, O.C., Chang, C.T. and Bettess, P. (1980) Drained, undrained, consolidating and dynamic behaviour assumptions in soils. *Géotechnique*, Vol. 30, no. 4, pp. 385–395.

Index

Page numbers in *italics* refer to figures and those in **bold** refer to tables.

T - #0113 - 111024 - C334 - 234/156/16 - PB - 9780367706708 - Gloss Lamination